ALLE ZEIT WACH
1842

G. R. Burgio L. Å. Hanson A. G. Ugazio (Eds.)

Immunology of the Neonate

With 24 Figures and 28 Tables

Springer-Verlag Berlin Heidelberg New York
London Paris Tokyo

Professor, Dr. G. ROBERTO BURGIO
Universitá di Pavia
Clinica Pediatrica, I-27100 Pavia, Italy

Professor Dr. LARS ÅKE HANSON
University of Götebôrg
Institute of Clinical Microbiology
Guldhedsgatan 10, S-413 46 Göteborg, Sweden

Professor Dr. ALBERTO G. UGAZIO
University of Brescia
Clinica Pediatrica
Ospedale dei Bambini
Via del Medolo 2, I-25100 Brescia, Italy

ISBN-13: 978-3-642-71096-4 e-ISBN-13: 978-3-642-71094-0
DOI: 10.1007/978-3-642-71094-0

Library of Congress Cataloging-in-Publication Data.
Immunology of the neonate.
Bibliography: p. Includes index.
1. Infants (Newborn)--Diseases--Immunological aspects--Congresses. 2. Immunology, Developmental--Congresses. I. Burgio, G. R. II. Hanson, Lars Å. III. Ugazio, A. G. (Alberto G.), 1944-
RJ255.I46 1986 618.92'01 86-17671
ISBN 0-387-16339-5 (U.S.)

Softcover reprint of the hardcover 1st edition 1987

2123/3145-543210

Preface

Over the last few years, many new observations have profoundly changed our concepts of the immune competence of the newborn. For the immune system, as for other systems and functions, the neonatal age represents a crucial transition period. In fact the immune characteristics of the fetus are likely to result from – or be conditioned by – several often contradictory physiological requirements. On the one hand, it would certainly be an advantage for the fetus to acquire a complete immunocompetence as soon as possible in order to be able to cope with the eventual transplacental passage of pathogenic microorganisms and possibly also in order to reject maternal cells occasionally crossing the placental barrier. This is actually what occurs, at least in part, during the 2nd and 3rd month of gestation when the fetus begins to acquire his biological individuality and at the same time the role of a "biological ego" resulting from the attainment by the immune system of the capacity to discriminate between self and nonself.

However, other physiological requirements would be satisfied only by a slowing down of the immune development and maturation. In fact, an early acquisition of a complete immune competence could possibly endanger the delicate immunological equilibrium between the fetal graft and the mother as a host. The result of these contradictory requirements is a complicated "immunological compromise" comprising the incomplete maturation of several cellular subsets as well as the active suppression of other cellular subsets and functions which are already fully developed.

During the neonatal period, physiological requirements change suddenly: the fetus is no longer a graft, and the danger of infections increases dramatically. The high susceptibility to infections of the neonate is likely to result, at least in part, from the persistence of the immune suppression or incomplete maturation of what, a few hours before, was a "fetal graft." The progressive understanding of these mechanisms is certainly improving our capacity to deal with the high susceptibility of the newborn to infections. On the other hand, the rapid transition from the fetal to the postnatal immunological trim makes the study of neonatal immunology extremely useful; this is therefore a unique model to investigate the regulatory mechanisms of the immune system as well as immune ontogeny. Furthermore, the intense neonatal traffic of various cellular subsets offers a unique opportunity to study the phenotype and function of cells that are not found easily later.

Scientists working in the field of neonatal immunology and in related areas of pediatrics and immunology had the opportunity of discussing many of these problems in the seventeenth-century setting of the *Almo Collegio Borromeo* in Pavia, certainly one of the oldest and most glorious university colleges in the world.

The contributions reported in this volume are the written testimony of the meeting. As happens in vital fields of research, certainly more questions have been raised than solved: new data and new problems contribute equally to the interest of the book.

We are much indebted to Rita Maccario and Alberto Martini for their invaluable help in the scientific organization of the meeting.

Pavia, July 1986 G. R. Burgio

Table of Contents

List of Contributors

The names of the senior authors are printed in semibold.

Immunologic Development

Introductory Remarks

ROBERT A. GOOD

801 6TH Street Souree, PO Box 31020, St. Petersburg, FL 33731-8920, USA

Immunologic development places critical focus on the neonatal and premature periods. Development of the cellular bases of immunologic processes is ongoing all through life. It is, of course, only one component of the constantly developing hematologic system. The constant development of hematologic and immunological cells must continue throughout life if individuality is to be preserved and if the bodily defense against all manner of potentially invasive organisms is to be effective and the tasks of carrying oxygen are to be fulfilled.

It is characteristic of all mammals, and in fact of all vertebrates, that T cells, B cells, phagocytic cells, and probably NK cells as well, all have a rich phylogenetic, as well as ontogenetic, developmental history. These histories are featured by the spectacular unfolding of genetic potential. This unfolding of genetic potential occurs in precise sequential stages, and each developmental change in the cellular sequence is now known to be under almost incredible genetic control.

Specificity of both the T-cell-mediated cellular immunities and the B-cell-mediated humoral immunities is rapidly coming to be understood as a consequence of regular rearrangement of genes located along chromosomes 14, 2, and 22. These are the genes which are responsible for the many V-region sequences that may be selected to achieve different specificities of the heavy and light Ig chains. On chromosome 14 are located not only the V-region, but a selection of D and J genes as well; and the rearrangements and choices from among these three sets account for specificities toward antigens. Also a selection of possible gene sequences that determines heavy chain constant region genes is available that accounts for the distributions of the Ig molecules.

This relatively new information concerning gene rearrangements goes a long way toward explaining the miracle of immunology posed long ago by Sir Peter Medewar. Medewar pointed out that the myriad specificities of antibody made it certain that a baby yet unborn could make antibody to an antigen that had not yet been synthesized.

No less impressive has been the unraveling of the molecular bases for specificities of the cell-mediated immunities including delayed allergy, allograft recognition and specific rejection, granuloma formation, and effective defense against many viruses, fungi, intracellular and facultatively intracellular bacterial pathogens, helminths, and protozoa. Here a regular rearrangement of genes deployed along both chromosome 14, and chromosome 7, which is taking place continuously during development of the T cells, accounts for the specificity of

Immunology of the Neonate
Burgio, Hanson, Ugazio (Eds.)

their alpha, beta, and perhaps gamma chain receptors. Genes for gamma are rearranged in all T cells, even though the function of these peptides is not yet known. Furthermore, the choice of which of the genes that determine the different heavy chain constant regions during development of the B cells and the selection of an appropriate light chain partner which will be put together with each of the heavy chains used is a most impressive and orderly process, which takes place during B cell differentiation. Understanding of the basis for these changes is making it possible to understand the mechanisms of recognition of antigenic specificity and also supplying the basis for determination of the distribution of the immunoglobulin molecules and their deployment in the body, e. g., their relation to functions of the K cells.

As if these complexities were not sufficient, the specific immunologic determinants must develop in a context in which they are restricted and their development is limited by the dictates of interactions with the other components of the supergene family defined in recent years (Early et al. 1980; Early and Hood 1981; Sin et al. 1984). The latter includes the immunoglobulins, the T cell receptors, the MHC determinants, and probably also certain T cell, e. g., CD3, or B cell developmental surface markers.

Similarly, crucial genes, molecular surface components, and developmental cellular processes which permit initial distinctions of self from non-self are beginning to be appreciated. The contribution of mechanisms for establishing self and non-self distinctions, achieving activation and expansion of cell populations to account for the effectiveness of the body defenses, is being elucidated. Powerful new biochemical and molecular biologic tools are increasingly making it possible to examine the orderly progression of events and precise play-out of the influence of crucial genetic events in a context of cellular activation, gene rearrangements, surface membrane antigenic displays, growth factors, differentiation factors, stop signals, and many others. Thus, we can look at critical issues of developmental immunology and hematology in terms that are rapidly expanding our understanding of embryonic processes. For example, thymic influences, thymic hormone actions, self-non-self recognition, cellular specificity repertoires, clonal expansion of cell populations, influences in production of a few crucial growth factors, differentiation factors, and many more must be considered.

As if all of this did not provide sufficient evidence of elaborate controls of immunity functions, it is now becoming urgently clear that another sleeping giant has arisen and a whole new science which considers and speaks about interaction of the major bodily networks must be taken into account. Thus, evidence of a powerful psychoneuroimmunobiology is at hand, and crucial interactions between brain, central nervous system functions, neurohumoral mediators, endocrinological influences, and the development and function of the humoral and cellular immunologic networks must be considered. Thus, a whole new endocrinology addressing immunologic function is at hand, analogous to the new endocrinology that influences all body functions and according to which every cell of the body must be looked upon as a endocrine producer and every cell as a target of endocrine action. Understanding of the immunologic developments and functions thus involves consideration of these several different influences if we are to understand how things work and how our bodies are defended (Tonegawa 1983). Thus, the immune system emerges as a constantly developing network of interacting cells and molecules capable of recognizing self from non-self and of eliminating or destroying that which is not self. This network functions to maintain the integrity of the individual in an environment which constantly threatens that integrity. The threat to the integrity of the individual, of course, may come from within in the form of cancer or as a consequence of the capacity of immunologic and nonspecific humoral or cellular events to damage not only foreigners but also the body itself. Thus, abnormalities of autoregulation of immunity functions, specific negative adaptations, or down-regulations, nonspecific mechanisms for destroying foreigners, all with potential for self-injury, are all being investigated intensively as bases of disease. Like the immune system itself, these mechanisms must be understood in developmental terms, genetic terms, and precise molecular terms if disease is to be understood and controlled and developmental disturbances recognized and corrected.

The influences of endorphins, enkephalins, insulin, adrenergic influences, thyroid and adrenal hormones, and other neurohumoral mediators are now recognized and have to be taken into account. Each can exert powerful influences on the immunologic functions and on their development and control (Ader 1981; Faumann 1985).

Thus, studies of the immunodeficiencies and analysis of developmental stages in immunology, such as the features that characterize premature infants and newborns, must emerge from the relatively superficial and descriptive analyses of the past. These must now be viewed in the seemingly complex yet simplifying powerful perspective of developmental molecular genetics and in the context of essential positive and negative interactions of the body's major networks.

Perturbations and disorders of development or abnormalities of genetic determination must be analyzed in the context of abnormal or defective responses, or defective and excessive cell-cell interactions, the genetic determinants of these, and the interactions of all of these, with the several components of the three major networks of the body. They must also be analyzed in terms of cell expansion mechanisms (growth factors) and peptides that control cell differentiation. We must consider anew the influences exerted by crucial minerals, trace metals, even calories, on these ongoing developmental and functional processes. From recent studies we must now even seek to understand the influences exerted by regular exercise on these functional processes in molecular terms.

As we pursue understanding of the lymphoid systems in these new perspectives, health and disease will be understood much better and new posibilities for maintenance of health, reduction of disease, and even prolongation of life and health seem probable.

In the context of the present symposium on the immunology of newborns and premature infants these new perspectives offer immense opportunities for analyzing and ultimately understanding the particular demands represented by transfer of the individual from the highly protected environment of the mother's womb into the hostile sea of microorganisms - bacteria, viruses, fungi, protozoa, and helminths - that have been chosen genetically as our ecological niche. The particular deficiencies of the bodily defenses of prematures and neonates have heretofore not been well understood. Indeed, they have not even been well described up to the present time. I, for one, do not think we yet know why the fetal allograft is not rejected by the mother for a 10-month period.

However, we can anticipate that the deliberations at this symposium in the magnificent and inspiring setting of the University of Pavia will contribute toward better understanding and management of the special immunologic and defense problems of the neonate in the years ahead.

References

Ader R (1981) Psychoneuroimmunology. Academic, New York

Early P, Huang H, Davis M, Calame K, Hood L (1980) An immunoglobulin heavy chain variable region gene is generated from three segments of DNA: V_H, D and J_H. Cell 19:981–992

Early P, Hood L (1981) Mouse immunoglobulin genes. In: SetPow JK, Hollaender A (eds) Genetic, engineering. Plenum, New York, pp 157–188

Fauman MA (1985) Psychiatric and neural aspects of allergy and immunology. In: Stone J (ed) Dermatologic immunology and allergy. St. Louis

Siu G, Clark S, Yoshikai Y, Malissen M, Yanagi Y, Strauss E, Mak T, Hood L (1984) The human T cell antigen receptor is encoded by variable, diversity, and joining gene segments that rearrange to generate a complete V gene. Cell 37:393–401

Tonegawa S (1983) Somatic generation of antibody diversity. Nature 302:575–581

T-Cell Development

O. Stutman

Immunology Program, Memorial Sloan-Kettering Cancer Center, New York, NY 10021, USA

1 Introduction

The study of ontogeny of T cells or of any other function has three aspects: the descriptive timing of appearance since inception, the quantitative evaluation of cell numbers, and the functional studies to detect if early functions are qualitatively different from those of the adult (Stutman 1985). The T-cell "deficiency" during the perinatal period has been explained as a consequence of: (a) simple quantitative deficits of the various components involved in the response tested, which may or may not include T cells; (b) some qualitative intrinsic properties of the responding T cells; (c) some peculiar excess of down-regulating influences, and (d) various combinations of the above (Stutman 1985). For a given function, the deficiency is usually multifactorial (see Sect. 4). Although the actual comparisons between species have problems, there is agreement that the onset of immunological functions follows the same sequence in all species, as lymphocytes appear during ontogeny and this is independent of the gestation time or life span of a given species (Solomon 1978). Due to space limitations, I will address only three related aspects of T-cell ontogeny: What is the consensus on T-cell development? What does the thymus export? and What is the nature of postthymic development?

2 Consensus Facts on T-Cell Development

Consensus means accord, and the present text is more a guideline for opinionated consensus rather than a list of agreements.

In mammals and birds, the first anatomical site of appearance of recognizable T cells is within the thymus (LeDouarin 1977; Stutman and Calkins 1977; Owen and Jenkinson 1981; Stutman 1985). Table 1 shows the timing of hemopoiesis, appearance of the thymic stroma, lymphopoiesis, and T-cell appearance during embryonic development in mice and humans (Stutman 1985). These three events mirror the three stages of T-cell development, i. e., prethymic, intrathymic, and postthymic (Stutman 1977, 1978).

Immunology of the Neonate
Burgio, Hanson, Ugazio (Eds.)

Table 1. Development of hemopoiesis and lymphopoiesis in mice and humans

Event	Time of appearance in gestation	
	Mouse	Human
Gestation period (days)	19–21	270–280
Hemopoiesis:		
Yolk sac, etc.	6–8	21– 28
Liver	10–12	42– 45
Spleen	15–16	90–100
Bone marrow	17–18	140–150
Thymus non-lymphoid *anlage*	10–11	40– 42
Lymphocytes:		
Thymus	12–14	70– 72
Thymus (T)	14–16	70– 72
Thymus (Cortex-Medulla)	17–18	70– 75
Thymus (TCR)	17	?
Thymus (IL-2 R)	16–17	?
Liver (B)	14–16	65– 68
Spleen	16–18	140–150
Spleen (B)	17–19	140–150
Spleen (T)	18–19	
Lymph nodes	18–19	70–130
Blood	17–19	120–130
Peyer's patches	19-Birth	120?

Hemopoiesis and *thymus anlage* are self-explanatory.
Lymphocytes implies the first appearance of cells with lymphoid morphology as well as appearance of T (defined by markers such as Thy 1,Ly in mice) and B cells (defined by Ig in surface or cytoplasm). TCR, T cell receptor, implying either gene rearrangement or expression of heterodimer; IL-2 R, IL-2 receptor. For references see text and Stutman (1985)

2.1 Extrathymic Maturation

"Extrathymic" is usually applied to events related to T cells in periphery. These include both postthymic and prethymic cells (cells which have not been within the intrathymic environment and are studied in vitro in the presence of T-cell products such as IL-2 or crude supernatants of activated T cells; see Sect. 4).

2.2 Prethymic Maturation

"Prethymic" is used to denote the events in hemopoietic tissues related to stem cell generation and possible commitment to the T lineage, and the events leading to homing and traffic of hemopoietic cells to the thymus. All the lymphoid cells within the thymus, including the first cohort during embryonic development, are of extrinsic origin and produced by colonization of the thymic stroma with cells of hemopoietic origin (Ford 1966; Micklem et al. 1966; Moore and Owen 1967;

LeDouarin 1977; Stutman 1972, 1976; Jotereau and LeDouarin 1982). The large basophilic cells of hemopoietic origin appearing in the thymus by days 11–12 of gestation are considered to be lymphoid stem cells (Moore and Owen 1967; Fontaine-Perus et al. 1981). Not much is known about such prethymic stem cells, apart from their extrinsic origin. Is thymic migration of stem cells random or predetermined extrathymically? Are they derived from stem cells with pluripotential or lineage restricted capabilities?

The actual embryonic site which produces the hemopietic cells that populate the thymus and hemopoietic tissues is still under debate: both the extraembryonic yolk sac (Moore and Owen 1967) and intraembryonic sites (Martin et al. 1978) are candidates for the source of stem cells colonizing the thymus.

Whether the initial site is extra- or intraembryonic, the primeval stem cells are incapable of direct migration to the tyhmus and have to colonize embryonic liver to initiate thymic colonization by liver-derived thymic stem cells (Stutman 1976). The capacity for direct migration to thymus grafts in nonirradiated murine models can be detected in embryonic liver and blood from day 14-15 of gestation (Stutman and Good 1971). Embryonic-derived cells have been shown to give rise to functional T cells in periphery after thymus traffic (Stutman and Good 1971; Stutman 1972, 1976).

The "dull-Ly 1" (Fowlkes and Mathieson 1985), "double negative Ly 2-/L3T4-" (Ceredig and MacDonald 1985), "Thy 1- Pgp-1+" (Trowbridge et al. 1985), and the "double negative B2A2- Thy 1-" thymocyte minorities (Scollay and Shortman 1985) have been termed "early thymocytes" and may represent true intrathymic T-cell progenitors capable of giving rise to all thymocyte subsets (see Sect. 2.3).

The term "precursor" is used rather loosely in present day immunology (Stutman 1985). Following Lajtha (1966) we proposed the use of precursor to define postthymic cells which gives rise to effector T cells and "progenitor" for the T cells giving rise to precursors (Miller and Stutman 1984). Unfortunately, in other publications I used the term "precursor" when "progenitor" was more appropriate (Stutman 1975). "Restricted stem cells" for myeloid and lymphoid (T and B) have been defined using radiation chimeras (Abramson et al. 1977). "Progenitors" as precursor predecessors have been defined in hemopoiesis (Lajtha 1966). In a strict sense, "progenitor" is the ultimate source of a given cell population (Lajtha 1966), while "stem cell" implies a self-maintaining population capable of giving rise to other cell lineages and cell types, including progenitors (Lajtha 1966). Thus, a cell which is committed to the lymphoid lineage could be termed a "lymphoid stem cell," while a cell which can give rise to differentiated progeny of a single lineage would be a "progenitor." Stem cells and progenitors are the generative compartments of a given lineage (Lajtha 1966).

A stage of prethymic commitment to T lineage has been suggested when a fraction of marrow cells was induced to express T-cell-related surface antigens, after short-term in vitro exposure to a variety of cyclic-AMP increasing compounds, which also included some putative thymic hormones (Komuro and Boyse 1973; Scheid et al. 1973, 1978). The in vitro "maturation" in these assays produces a cell with the characteristics of a cortical thymocyte (Scheid et al. 1973) and not of a peripheral T cell (Stutman 1977, 1978, 1983). Such "inducible" cells have been termed "prothymocytes" (Komuro and Boyse 1973) and are immunologically nonfunctional after induction (Scheid et al. 1973; Stutman 1977, 1983); are thymus independent for development (detected in athymic nude mice spleen and marrow, Komuro and Boyse 1973); appear early during ontogeny (detected in the 14-day old embryonic liver, Kumoro and Boyse 1973); and contain the cells which can repopulate the thymus in lethally irradiated mice, 2 or more weeks after cell injection (Komuro et al. 1975; Kadish and Basch 1976; Basch et al. 1978; Greiner et al. 1982). "Thymus-homing" marrow cells have been also defined using a short-term assay in which fluorescein-labelled cells injected into lethally irradiated mice are detected in the thymus 3–24 h after injection (Lepault et al. 1983); these cells, termed "prethymic," are probably similar to prothymocytes. Prothymocytes and marrow thymus-homing cells are Thy 1, Tla, Ly 1, and Ly 2 negative in mice (Komuro and Boyse 1973; Scheid et al. 1973, 1978; Basch et al. 1978; Lepault et al. 1983) although they can express such antigens after either in vitro induction (Komuro and Boyse 1973) or after in vivo homing to the thymus (Lepault et al. 1983). However, the thymus-migrating prothymocytes are included within the murine marrow cells which express low levels of Thy 1 (Basch and Berman 1982) and in rat marrow are clearly Thy 1 positive (Greiner et al. 1982).

The in vitro short-term induction triggers the simultaneous expression of all the markers tested (Komuro and Boyse 1973; Scheid et al. 1973, 1978), while in the short-term in vivo experiments the thymus-homing cells do not express Tla and Lyt expression precedes Thy 1 (Lepault et al. 1983; see Sect. 2.3).

A problem with the in vitro induction assays is the high frequency of prothymocytes (5%–20%) in marrow (Komuro and Boyse 1973; Scheid et al. 1973, 1978). The enrichment of thymus-migrating prothymocytes obtained by different procedures in mice and rats, with maximum values of 0.25% of marrow cells having prothymocyte thymus-migrating capacity (Basch et al. 1978; Greiner et al. 1982), fits much better with other progenitor frequencies for hemopoiesis (Metcalf 1977). Thus, not all of the in vitro-induced prothymocytes have thymus-migrating capacity, or alternatively, not all the induced cells are prothymocytes.

Time-gaps between detection of donor cell and cells with T-cell markers have been observed in thymus of adult radiation chimeras (Micklem et al. 1966; Kadish and Basch 1976). This contrasts with the detection of Thy 1+ in the thymus-homing population in the thymus, 20–24 h after injection of marrow cells to irradiated hosts (Lepault et al. 1983).

It is possible that initial seeding, measured in the short-term assays (Lepault et al. 1983), may not define the subsequent proliferation, selection, and differentiation within thymus, especially since thymus seems to play an active role in allowing certain hemopoietic cells to proliferate within its stroma, as is the case with syngeneic preference (SP). SP is the preference for intrathymic proliferation for the most histocompatible cells when mixtures of syngeneic and allogeneic hemopoietic cells are used (Stutman 1977, 1978). SP is mediated by an intrinsic mechanism within thymus that can be reversibly overcome by irradiation with 750 rads (Stutman 1978). In birds, the entrance of hemopoietic cells to the thymus during embryonic and perinatal life is cyclic, with short periods of active immigration lasting approximately 1 day, followed by 5-day-long refractory periods (Jotereau and LeDouarin 1982). It is still unknown whether such cycles

Table 2. Effect of age on thymic export: appearance of F+ cells in tissues of intrathymically F injected CBA/H mice of different ages

Age of mice (days)	Number of F+ cells/10^5 cells/3 h			
	Blood	Spleen	Lymph nodes	Marrow
1	78 ± 28	4 ± 0.3	190 ± 19[a]	ND[b]
3	191 ± 41	9 ± 1.8	160 ± 23[a]	ND[b]
6	143 ± 38	3 ± 0.3	59 ± 5.2	9 ± 3.6
30	265 ± 40	31 ± 3.5	22 ± 3.3	1 ± 0.5
120	60 ± 19	3 ± 0.4	2 ± 0.4	9 ± 1.6
180	11 ± 12	3 ± 0.6	1 ± 0.3	1 ± 0.6
360	10 ± 9	2 ± 0.6	1 ± 0.4	0

Mean ± SEM of values for individual CBA/H mice injected intrathymically (IT) with fluorescein isothiocyanate (F) at indicated ages; 4–10 mice per point; 3 h after IT injection

[a] Pooled lymph nodes from 8 and 6 mice

[b] ND, not done

occur beyond the early perinatal stages and whether they also occur in mammals. Table 2 shows that thymic export is not constant throughout life (see Sect. 3).

Low numbers of stem cells are responsible for thymus repopulation in radiation chimeras (Micklem et al. 1972; Wallis et al. 1975; Ezine et al. 1984) and in nonirradiated parabiosis partners (Ford 1966), suggesting a slow rate of migration of stem cells to the thymus. Some thymus repopulation studies suggest the development of distinct clones within thymus (Ezine et al. 1984). If indeed the rate of entry into the thymus is low and acceptance by the thymus is active (Stutman 1978) and cyclic (Jotereau and LeDouarin 1982), the stem cells or progenitor cells entering the thymus must have considerable proliferative capacity to be able to produce large numbers of thymocytes, most of which are short lived (Scollay and Shortman 1983).

One open question concerning the prethymic compartment in marrow (prothymocytes, thymus-homing cells, "lymphoid" stem cells, etc.) is their degree of differentiation along the T pathway. One study suggests that the thymus-homing marrow cells express anti-H2 receptors prethymically (Morrissey et al. 1982). The fact that marrow or embryonic liver in humans and mice have the genes for chains of the T receptor not rearranged (Yoshikai et al. 1984; Snodgrass et al. 1985; Born et al. 1985) speaks against this interpretation. Thus, if prethymic "commitment" to T lineage does exist, it does not include the rearrangement of genes coding for the T-cell receptor.

2.3 Intrathymic Maturation

"Intrathymic" is self explanatory. The issue of thymic hormones, whether produced in the thymus or elsewhere, is beyond the scope of this review (Stutman 1983). The intrathymic stage of T-cell development comprises at least four integrated events: 1. irreversible commitment of hemopoietic cells towards T-cell differentiation; 2. "selection" of the T-cell repertoire which provides a bias towards self-MHC antigens; 3. selection of populations for export; and 4. functional diversification into subsets (Stutman 1978, 1985). An intrinsic part of intrathymic differentiation is the initiation of the gene rearrangements necessary for the expression of T-cell receptors (Davis et al. 1984) and the definition of the individual repertoire in relation to class I and class II self-antigens (Stutman 1985). This sequence of events applies both to the first cohort of thymocytes produced during early ontogeny and to the development of new thymocytes in the adult.

Thymus is the first site where T-cell receptor (TCR)-related gene rearrangements are expressed during ontogeny. By day 15 of gestation the genes for the beta chain of TCR are in germline configuration, and rearrangements begin to accumulate by day 17 in murine thymocytes (Snodgrass et al. 1985; Raulet et al. 1985; Born et al. 1985). A detectable alpha chain of TCR appears later, approximatley at birth (Raulet et al. 1985). Precipitable heterodimers are detected on thymocytes at 17 days of age (Snodgrass et al. 1985); however, such dimers could be either beta-gamma or beta-alpha, the latter being the adult type (Raulet et al. 1985). A monoclonal antibody detecting a beta chain determinant of TCR (KJ16) shows the first positive cells in the murine thymus at 17–18 days of gestation (Rohem et al. 1984). Thus, the first cohorts of Thy 1 +/Ly + thymocytes which appear before day 17–18 of gestation do not express TCR-related structures yet. Only 15% of thymocytes are KJ16+ in the adult (Rohem et al. 1984). In humans, TCR is linked for its surface expression to T3 (Reinherz et al. 1982). The frequency of T3+ thymocytes in thymi ranging from 2 months to 2 years of age is approximately 10% (Reinherz et al. 1980). Similarly, the expression of TCR in thymocytes of the same age as above is limited to a very small fraction of cells (Acuto et al. 1983).

Studies with early thymocytes of the "dull Ly 1" type (dLy1) in adult thymus (Samelson et al. 1985) showed that the expression of beta chain of TCR and of the murine equivalent of the T3-delta chain (van den Elsen et al. 1985) are early events in maturation and are followed by alpha chain (as is the case in ontogeny, Raulet et al. 1985). Moreover, the formation of T-T hybrids with dLy1 cells showed two patterns: hybrids with germline TCR genes (pro-T) and hybrids with rearranged beta chain of TCR only (pre-T). The phenotype with beta and alpha rearranged giving rise to complete TCR is a later stage of development. Some early thymocytes in the adult thymus, such as those defined by absence or low expresion of Thy 1, have no rearranged beta-chain genes of TCR (Trowbridge et al. 1985). The "double negative" thymocytes (Ceredig and MacDonald 1985) express good levels of beta chain mRNA and also gamma chain, while the more mature thymocyte subsets express beta and increasing amounts of alpha chain mRNA (Raulet et al. 1985). The replacement of gamma for alpha appears as a maturation event in TCR development (Raulet et al. 1985). Rearrangement and expression of beta appears as the earliest event related to TCR maturation, both during early ontogeny and during T-cell differentiation in the adult (see above). Some of these thymocyte subpopulations appear as true intrathymic progenitors.

Older studies have shown antigen binding cells in the murine embryonic thymus as early as day 14 of embryonation, which reach a peak by day 17 and decrease to adult low levels at birth (Dwyer and MacKay 1972a; Decker et al. 1974). In humans, antigen-binding cells appear in thymus at 77–84 days of embryonation (Dwyer and MacKay 1972b; Stites et al. 1972). Thus, it seems that antigen-binding cells appear before recognizable T cells (with detectable T-related surface antigens or with expressed TCR) are found in the thymus of mice and humans. The question on the reliability of antigen-binding assays is beyond the scope of this review.

Based on the thymus structure, the highly controlled excess production of lymphocytes with subsequent selective reduction, and the experimental data showing that T-cell development (and function) implied cell populations interacting in a highly predictable manner, we proposed a model describing the concordant development of *cell pairs* (Stutman 1978). The intrathymic unit of differentiation is a pair of cells, one of which is the prethymocyte, which may give rise to functional T cells, and the other is the "inducer" cell, which is part of the thymic nonlymphoid stroma and provides signals related to differentiation and repertoire (Stutman 1978). Such paired interactions also take place in the periphery (Stutman 1978; Ching and Miller 1981). Thus, ontogeny would imply positive selection of concordant matches between the prethymocyte and its set of "inducer" partners (Stutman 1978). The nonconcordant matches would produce nonfunctional differentiation and be negatively selected intrathymycally (Stutman 1978), and may represent the dead-end cortical thymocytes of the Scollay and Shortman (1983) model. Escape of nonconcordant matches could be involved in malignant transformation (Stutman 1978).

There are abundant studies of thymocyte subpopulations during ontogeny and during adult T maturation (Kamarck and Gottlieb 1977; Mathieson et al. 1981; van Ewijk et al. 1982; Scollay and Shortman 1983, 1985; Scollay et al. 1984a, c; Ceredig et al. 1983a–c; Ceredig and MacDonald 1985; Mathieson and Fowlkes 1984; Fowlkes and Mathieson 1985). The following conclusions have been reached from these studies:

1. No readily detectable T-cell-related antigens (such as Tla, Thy 1, Ly 1, Ly 2, Ly 3, L3, T4) are expressed within the murine thymus before day 14 of gestation. From day 14–15 onwards, and following the exponential growth of the lymphoid cells in thymus, there is a gradual increase in antigen positive cells.

2. The in vitro prothymocyte conversion assays (Sect. 2.2) and some of the earlier typing studies (Cantor and Boyse 1977; Kamarck and Gottlieb 1977) showed the simultaneous expression of all the T-cell-related antigens at some stage of intrathymic development. However, there is an asynchronous and/or sequential appearance of some antigens on thymocytes during ontogeny (Mathieson et al. 1981; Van Ewijk et al. 1982; Ceredig et al. 1983a–c). Ly 1 is detected at day 15–16 of embryonation as the main antigen on thymocytes (aproximately 50% of the thymocytes are positive), with only a minority expressing Ly 2; Ly 2+ cells become a majority population by day 17–18 (Mathieson et al. 1981; Van Ewijk et al. 1982; Ceredig et al. 1983a) coinciding with the appearance of L3T4+ cells (Ceredig et al. 1983a). The "adult" phenotype is reached approxiately at birth, and 90% of thymocytes from birth onward have the Ly 1+/Ly 2+/L3T4+ (Thy 1+) phenotype (Mathieson et al. 1981; Ceredig et al. 1983a; Scollay and Shortman 1983; Scollay et al. 1984a, c; Fowlkes and Mathieson 1985; Ceredig and

MacDonald 1985). A similar picture has been described in the human perinatal thymus (Reinherz et al. 1980; Schlossman et al. 1983). Although initially influenced strongly by the murine studies, the human model provided a new framework for the murine studies based on Ly 2/T8 and L3T4/T4 parallelisms (Ceredig et al. 1983a; Scollay et al. 1984a; Mathieson and Fowlkes 1984). At around birth, the murine thymus contains a minority of approximately 10% which express either Ly 2 or L3T4, as is the case in the human thymus (Reinherz et al. 1980).

3. The asynchronous appearance of thymocyte antigens would support the intrathymic development of more than one subclass T lineage (Scollay and Shortman 1983, 1985; Scollay et al. 1984a; Mathieson and Fowlkes 1984; Fowlkes and Mathieson 1985; Ceredig and MacDonald 1985), while the term "sequential" suggests adherence to single intrathymic lineage models. Most of the present information supports the view of either separate lineages or intrathymic diversification for the various T-cell subsets.

4. However, none of the generalizations concerning the interrelationships between thymocyte subsets is absolute. Firstly, embryonic Ly 2– thymocytes can give rise to 40%–80% Ly 2+ cells after culture; similar experiments with adult thymocytes give rise to only 6% Ly 2+ cells (Ceredig et al. 1983c). Secondly, dLy1 thymocytes can give rise both in vivo and in vitro to all the thymocyte subpopulations, acting as a true progenitor (Mathieson and Fowlkes 1984; Fowlkes and Mathieson 1985). Thus, Ly 2– cells can give rise to Ly 2+ cells, and dLy 1+ cells can give rise to Ly 2+ as well as Ly 1+ cells at a thymic level, in spite of the fact that such interconversion does not occur readily with postthymic subsets in periphery (Huber et al. 1976; Miller and Stutman 1982).

5. The actual functional significance of the 7-12 thymocyte subsets defined by multiparameter analysis using surface antigens and cell size (Scollay and Shortman 1983, 1985; Scollay et al. 1984a; Fowlkes and Mathieson 1985) is difficult to establish, even in adult mice. It is unclear whether some particular surface phenotype is indicative of separate lineages or is due to different stages within a lineage, and no clear precursor-effector relationships have been shown directly in most of these subset models (Stutman 1985).

6. The importance of some minority populations, ranging from 0.2% to 5% of the whole thymocyte population in the young adult thymus, representing the generative compartment for the functional T cells, is emerging and needs further study (Fowlkes and Mathieson 1985; Scollay and Shortman 1985). The "dead-end" thymic lineages, either destined to intrathymic death (Scollay and Shortman 1983; Scollay et al. 1984a) or as resident nonexported populations of steroid-resistant cells (Stutman 1977; Scollay and Shortman 1983), seem to be part of the selection process hypothesized above.

7. A good example of the problems with assuming cell properties based on surface phenotype is the case of RTM (recent thymic migrants, see Sect. 3) versus MEL-14++ thymocytes in mice (Reichert et al. 1984; Fink et al. 1985). Since RTM have the phenotype of medullary thymocytes (Scollay et al. 1984c) it was concluded that the medulla was the site where migrants accumulated prior to departure (Scollay and Shortman 1983; Scollay et al. 1984a, c). However, studies with a monoclonal antibody (MEL-14), which recognizes structures involved in lymphocyte homing to periphery, showed that the 1%–3% MEL-14++ thymocy-

tes, which could be candidates for migration, were located exlusively in the cortex and had a "mature" phenotype (Reichert et al. 1984). However, detailed multiparameter cytofluorographic analysis has been unable to detect any functionally mature cells of "medullary" phenotype in the thymus cortex (Shortman et al. 1985). Clusters of T3+ cells (which belong to the mature subpopulations of medullary thymocytes; Reinherz and Schlossman 1980) have been detected in the human thymic cortex (Bhan et al. 1980). Another complication is that some of the most immature thymocytes, such as the double negative Ly 2-/L3T4- subpopulation, are also MEL-14++ (Ceredig and MacDonald 1985). Would this imply that such immature cells are also ready for export?

8. Studies on the ontogeny of the venules with high endothelium have shown that such structures appear after birth in mice, and even at 6 days of age venules in the lymph nodes and Peyer's patches are still underdeveloped (Miller 1969). Thus, during ontogeny, high endothelium venules (which have the binding sites for MEL-14+ T cells) appear in peripheral lymphoid organs after the beginning of thymic export. Therefore, lymphocyte traffic during early ontogeny takes place via still undefined vascular sites.

9. The actual delineation of cortex and medulla in the murine thymus begins et day 17–18 of embryonation (Mandel and Russell 1971; Biggar et al. 1972) and at days 84–140 in humans (Haynes 1984).

This list of "riddles" could go on, and it may be questioned why they are included in a "consensus." However, they are good examples of the problems faced with the static analysis of important minority populations and the dangers of considering any marker or property as an absolute parameter related to cell function (Stutman 1985).

The thymic stroma is a mosaic of cell types containing epithelial cells of endodermal and ectodermal origin and mesenchyme of neural crest origin (LeDouarin 1977; Jenkinson et al. 1981; Kendall 1981). Macrophages and some dendritic cells are derived from blood-borne stem cells (Longo and Schwartz 1980; Barclay and Mayrohofer 1981). Thus, a component of the nonlymphoid stroma, which includes some of the dendritic cells expressing class II MHC antigens, is also "extrinsic." Class II MHC antigens are expressed throughout the adult thymic cortex on dendritic cells, which include both epithelial and mesenchymal interdigitating cells (Rouse et al. 1979). Class I antigens are expressed in a variable pattern in the cortex, and both class I and II antigens are expressed on the medullary stroma and large MHC-negative areas are found in the corticomedullary regions (Rouse et al. 1979; Van Ewijk et al. 1980). During ontogeny, epithelial cells expressing both class I and class II antigens are detected from the 14th day of gestation, while the mesenchymal cells express only class I antigens beginning on day 15 (Jenkinson et al. 1981). In summary, expression of class I and class II MHC antigens in the thymic nonlymphoid stroma occurs early during ontogeny and with a peculiar microenvironmental distribution within thymus. A similar picture has been described in human thymus with monoclonal antibodies which recognize epithelial and mesodermal stroma (Haynes 1984) as well as with expression of class I and class II MHC antigens by stromal cells (Janossy 1983; Haynes 1984). One important and still open question is which of the different stromal elements in the thymus is involved in conferring to the developing thymocytes the capacity to recognize self-class I and class II MHC antigens.

The proliferation of murine thymocytes in mixed lymphocyte cultures with syngeneic adult non-T cells was observed in mice by Howe et al. (1970) and was considered as a special type of the syngeneic mixed lymphocyte response (Battisto and Ponzio 1981). The murine SMLR, whether in thymus or other lymphoid organs, is mediated by Ly 1+ L3T4 + Thy 1+ Ly 2– T cells which respond to self class II on stimulatory cells of the macrophage-dendritic lineage and produce IL-2 and other lymphokines (Lattime et al. 1980, 1981, 1982, 1983; Stutman et al. 1980). The capacity to respond in SMLR by proliferation or IL-2 production appears in thymus at days 17–18 of embryonic age, peaks at around birth, and disappears at days 14–21 of postnatal age (Howe et al. 1970; Lattime et al. 1982, 1983). Once the self class-II-triggered IL-2 production by Lyt 1+ L3T4+ T cells develops in the periphery, it is capable of extensive self-renewal in the absence of the tyhmus, since it is detected 9–12 months after adult thymectomy (Stutman et al. 1980; Lattime et al. 1982; see Sects. 3 and 4). From all the properties of class II self-recognition, we postulated that it could represent the physiological signaling for IL-2 production, independent of nominal exogenous antigen (Stutman et al. 1980, 1982; Lattime et al. 1981, 1982, 1983). The compartmentalization of class II+ stromal elements within the thymus and peripheral lymphoid tissues in mice and humans favors local microenvironments where Lyt 1+ L3T4+ T lymphocytes get in contact with class II+ stromal cells, with in situ production of IL-2 (Stutman et al. 1980, 1982). This morphological pairing interaction supports both our views on concordant development of pairs of cells as well as the concept of the "Lyt 1/L3T4-self Ia recognizing -IL-2 producing" lineage (Stutman et al. 1980, 1982). Its function is to produce low levels of IL-2 independently of exogenous antigenic stimulation. The peculiar environments in the lymphoid tissues where class II+ dendritic cells are abundant (Janossy 1983) could ensure local micro-circuits where the lymphokine-producing short-range interaction takes place. These microcircuits are involved in intra- and postthymic differentiation (and the peripheral type of T-cell renewal), probably fixing differentiative changes through a round of cell division (Stutman et al. 1980, 1982). The nature of the differentiative signals (since IL-2 per se is necessary but not sufficient for differentiation) in this model is still undefined. The regulation of IL-2R expression in the differentiating cells (Stutman et al. 1980, 1982) could imply that T cells may express IL-2R at some stages of their development (Stutman 1983, 1985). Subsets of thymocytes of the immature types express IL-2R spontaneously (Ceredig et al. 1985; Raulet 1985).

The acquisition of bias towards self MHC determinants, which is the basis for associative recognition as well as MHC restriction, is one of the instructions imparted by the thymus, although there are no true ontogeny studies on the subject (Möller 1978; Wagner et al. 1982). However, two points are worth stressing. First, the repertoire of the intrathymic thymocytes is preferentially guided by the MHC of the thymic stroma, while the repertoire of the postthymic T cells can be amplified to include extrathymic MHC determinants (Zinkernagel 1978; Doherty et al. 1981; Bradley et al. 1982; Singer et al. 1982; Kruisbeek et al. 1983). Secondly, postthymic T cells in radiation chimeras show different patterns depending on the class of MHC restricting element: class II restricted responses have a tendency to predominantly express the intrathymic self-repertoire, while class I restricted responses express both the thymic and extrathymic host repertoire (Singer et al. 1982). In summary, regardless of the proposed mechanisms for intrathymic instruction, there is a strong posttyhmic influence which can modify the intrathymically defined repertoire in some murine models.

2.4 Posttyhmic Maturation

"Postthymic" is used to describe the events which occur in the peripheral lymphoid organs after export of the thymus-processed T cells (Stutman 1975, 1977, 1978). The two main issues of postthymic development, regarding what the thymus exports and the nature of posttyhmic development, will be addressed in Sects. 3 and 4.

3 What Does the Thymus Export?

Are thymus emigrants fully competent or do they require a phase of postthymic maturation? Although there is no complete agreement concerning the state of maturity of the thymic emigrants, there is agreement that it is arithmetically impossible to maintain the pool of peripheral T lymphocytes based exclusively on thymic export (Joel et al. 1974; Stutman et al. 1980; Scollay et al. 1980; Miller and Stutman 1984). If we accept that the murine adult thymus exports approximately $>10^6$ cells a day (Scollay et al. 1980; also supported by direct cannulation of venous output of the thymus in calfs, Joel et al. 1974) and that at least 30%–50% of the T-cell pool in periphery have a rapid turnover (Everett et al. 1964; Rocha et al. 1983), we must also accept that it is impossible to maintain a T-cell pool of $>10^9$ in periphery at constant homeostatic levels by thymus export alone. This is true especially since antigenic stimulation does not seem to influence thymus export (Scollay and Shortman 1984; our unpublished studies). Thus, even in putative conditions of "demand" in periphery, the cellular output of the thymus remains constant at a given age. However, thymic export is not constant and declines with age (Scollay et al. 1980; Table 2).

If we also accept that there is extensive expansion of the exported T cells in the postthymic compartment (Miller and Stutman 1984), the question can be asked: What is expanded in the periphery? Mature T cells driven predominantly by exogenous antigen as proposed by Smith (1984)? Some type of progenitors-precursors which differentiate into the mature effectors? Or both? The former view is supported mainly by Scollay (Scollay et al. 1984a, b) and other studies (Fink et al. 1985). The latter view(s) are supported mainly by our own studies (Stutman 1977, 1978, 1985; Stutman et al. 1980, 1982; Miller and Stutman 1982, 1984). The ontogeny studies by Piguet et al. (1981) as well as studies on repertoire development in chimeras would also support the concept of further maturation in periphery.

The notion of postthymic "precursors" (or PTP) cells in periphery was established by us out of the conceptual need to explain some of our experiments on the late restoration of neonatally thymectomized hosts (Stutman 1975, 1977) and as a symmetrical model to the schemes for hemopoietic maturation. The PTP model implied two concepts: 1. the existence in periphery of a recognizable progenitor compartment derived by thymus processing and export, and 2. the fact that the exported component also included the PTP or other cells which would mature further in periphery (Stutman 1975, 1977, 1978; Stutman et al. 1980, 1982). The main properties of the PTP cells were expression of Ly 1, 2, 3, and Thy 1 antigens, steroid sensitivity, nylon adherence, spleen homing, and no detectable immunological function (Stutman 1975, 1977). Such cells are capable of extensive expansion in T-depleted hosts in the presence of thymic influence (Stutman 1975). In 60-day-old neonatally thymectomized hosts, 95%–98% of the T cells responding to mitogens or alloantigens 40 days later were derived from the innoculum of 5×10^6 partially enriched PTP cells from newborn spleens (Stutman and Shen 1979). Even with the improbable assumption that "all" of the injected cells were PTP, they expanded by a factor of 20 giving rise to approximately 10^8 T cells under thymic influence (Stutman 1978; Stutman and Shen 1979). Thus, from a functional standpoint, PTP cells act as a generative compartment and are capable of extensive expansion and functional maturation.

The labelling of thymocytes with fluorescein isothiocyanate (F) by direct injection into the thymus allows the analysis of the cells leaving the thymus 3–4 h later (Scollay et al. 1980). Based on expression of Thy 1, H2 receptors for peanut agglutinin, Ly 1 and Ly 2, it was concluded that the recent thymic migrants (RTM) have a mature phenotype (Scollay 1982). However, immature RTM of a PTP type could be included in the 10%–15% of spleen homing Ly 1+2+ migrants, although the bulk of RTM had a mature phenotype (Scollay 1982). RTM have been detected also in bone marrow (Scollay et al. 1980; see Table 2), as is the case for PTP cells (Stutman 1975). The mature phenotype of RTM was comparable to that of medullary thymocytes (Scollay et al. 1984a, c; Scollay 1982, Scollay and Shortman 1983). However, both medullary thymocytes and RTM differ from peripheral T cells in size, buoyant density, and expression of the B2A2 antigen (Scollay et al. 1984c). Furthermore, RTM leave the thymus expressing a medullary phenotype and acquire the peripheral phenotype after arrival at the periphery (Scollay et al. 1984c). Two conclusions were reached: the possible medullary origin of RTM (suggested but not proven by the similarities between medullary thymocytes and RTM); and there are maturation steps that occur after arrival at the periphery even though the RTM leave the thymus with a mature phenotype by most criteria (Scollay et al. 1984c). Not all medullary thymocytes are functionally mature, and 20%–50% of medullary thymocytes are immunologically inactive in the same assays to be commented upon below (Chen et al. 1983). So far the PTP and RTM models are compatible. However, functional studies showed that RTM are comparable to peripheral T cells in their capacity to produce cytotoxic T cells and proliferation in response to mitogens (Scollay et al. 1984b). The conclusions reached were that RTM are fully competent a few hours after leaving the thymus and that there is no evidence of a postthymic precursor-type cell needing further maturation in periphery (Scollay et al. 1984b). However, the study has two intrinsic problems which make the conclusions highly questionable. Firstly, the in vitro assays used are long (7–9 days) and supplemented with a variety of products aiming at optimal plating efficiency and functional expression (such as 13% of a Con A supernatant from rat spleen cells or additional Con A) which may actually induce differentiation during the assay proper. Thus, one may say that what was measured was 7–9 days away from what emigrated from thymus and was grown in conditions which favor maturation and expansion of competent T cells. Secondly, RTM had a reduced plating efficiency of 20%–30% compared with 80%–100% for peripheral T cells. Whether the low plating efficiency of the enriched F+ RTM is due to actual loss of 70%–80% of the RTM cells or whether the assays are selecting for the 20%–30% mature T cells clonable in the type of assay used is irrelevant for the argument, since in either case only a fraction of the RTM would be tested. Thus, none of the two conclusions reached by Scollay et al. (1984b) concerning the absence of immature RTM is warranted.

In summary, the thymus seems to export an heterogeneous population of cells which includes a fraction of functionally mature cells (Scollay et al. 1984b) and other cell types, such as the "Ly 1/L3T4-self IA/E recognizing-IL 2 producing" lineage (see above) and PTP cells acting as progenitors in periphery, which are important in the generative aspects of the peripheral T-cell pool (Stutman 1975, 1977, 1978; Stutman and Shen 1979; Miller and Stutman 1982, 1984). However, the present prevalent view is that fully mature thymocytes are exported to periphery (Scollay et al. 1984a, b; Fowlkes and Mathieson 1985).

4 What Is the Nature of Postthymic Development?

A model for the enumeration of postthymic progenitor cells has been recently described by us (Miller and Stutman 1984). Since our estimates of the frequency of progenitors which can give rise to CTL precursors (pCTL) for a given H2 haplotype in periphery are 5.5–6.4 per 10^6 T cells, with 2.9–8.6 95% confidence limits (Miller and Stutman 1984), the daily output of RTM would contain but a few progenitors, based on the estimates discussed in Sect. 3. Furthermore, since the progenitor assay needs in vivo expansion to determine the actual number of pCTL derived from the injected posttyhmic progenitors, the in vitro precursor assays per se cannot detect such progenitor compartment (Miller and Stutman 1984). Especially, since in our model, each progenitor cell can give rise to approximatelly 10 000 allo-specific pCTL during the 12-week interval used, in an antigen independent manner (Miller and Stutman 1984). Conversely, the factor of expansion of the precursors for either CTL or IL-2-producing T cell in these animals was 5–6, when comparing precursor input on day 0 versus precursor frequency at 12 weeks (Miller and Stutman 1984). Thus, the precursor compartment has a more limited capacity for expansion than the progenitors when assayed against the same set of alloantigens (Miller and Stutman 1984). The progenitor assay makes two predictions, which were supported by the experimental data: First, the distribution of "positive" recipients as a function of T-cell number will follow a zero-order Poisson distribution (this allowed the calculation of progenitor frequencies); and secondly, mice receiving limiting numbers of progenitors will develop positive reactions for any H2 haplotype, independently of their response to any other haplotype (Miller and Stutman 1982, 1984). One interesting aspect of this model is that monoclonal mice capable of responding to a single H2 haplotype were detected in spite of the extensive expansion of the progenitors, which could favor mutation. Repertoire diversification via mutation has been detected in T-cell lines as well as in in vitro differentiation assays (Augustin and Sim 1984; Ching and Miller 1982). It is possible that the in vivo situation in our model may either select against a mutated repertoire or allow "clonal dominance" to occur. The fact that depletion of Thy 1+ cells from the T inoculum abrogated the generation of pCTL supports the view that the progenitors in adult splenic T cells are postthymic (Miller and Stutman 1984). These experiments demonstrate three points, which were previously supported by more indirect evidence (Stutman 1977, 1978): 1. the existence of postthymic progenitors in periphery which act as a generative compartment for the T cell pool; 2. the antigen-independent expansion and maturation of such compartment in periphery (Miller and Stutman 1982, 1984), and 3. the thymus independence of the peripheral expansion of the postthymic progenitors and precursors for allo-CTL and IL-2 producers (Miller and Stutman 1984).

Some studies have described the presence of immature T cells in periphery during early ontogeny in mice (Piguet et al. 1981; Stutman and Ishizaka 1982; Stutman et al. 1982) and humans (Yokoi et al. 1982; Maccario et al. 1981, 1983; Griffiths-Chu et al. 1984; Wilson 1985). By "immature" is meant cells with either immature phenotype or deficiencies in functional behavior or repertoire. The presence of such cells in periphery would indicate that, at least during ontogeny and early perinatal stages, the thymus may be exporting immature or partially differentiated T cells. Such immature T cells would also support the view that the deficiencies in function detected during early ontogeny could be not only quantitative as in Ceredig (1979) or Ishizaka and Stutman (1983) but also qualitative as in Piguet et al. (1981) or Stutman and Ishizaka (1982). Human cord blood contains T Cells of unusual phenotype which include PNA+/T8+ and T8+/T3- cells (Maccario et al. 1981, 1983), high frequency of T10+ (Wilson 1985) as well as some T6+ (Griffiths-Chu et al. 1984), T4+/T8+ (Griffiths-Chu et al. 1984), and T3+T6+ cells (Griffiths-Chu et al. 1984). T6+ cells were observed especially in preterm and antenatally stressed infants (Wilson 1985). There are no T6+ in adult peripheral blood T cells, T10+ are rare in resting blood T cells (Schlossman et al. 1983), and only 3% of adult blood T cells are T4+/T8+ (Blue et al. 1985). Thymocytes are 90% T4+T8+, 70% T6+, and 95% T10+ (Reinherz et al. 1980; Reinherz and Schlossman 1980; Schlossman et al. 1983). Based on surface phenotype and frequency of cells in DNA synthetic S phase, it was proposed that the Thy 1+ T cells in murine spleen and lymph nodes resembled cortical thymocytes rather than mature peripheral T cells (Piguet et al. 1981). Functionally, the 4- to 5-day-old T cells have low reactivity to mitogens, poor capacity to induce graft-versus-host reactions, poor capacity to proliferate in response to alloantigens and to help B cells in vivo or in vitro; however, they have good capacity to generate alloreactive CTL (Piguet et al. 1981), although they have impaired capacity to generate anti-self-TNP CTL, which mature later in some strains (Stutman and Ishizaka 1982). Six-day-old T cells in spleen and nodes have normal capacity to produce IL-2 on a per cell basis, but have lower numbers than adult, especially in spleen (Ishizaka and Stutman 1983). In the cord blood studies, it was observed that after activation only a low frequeny of T cells expressed class II antigens when compared with adult T cells, while the expression of IL-2R was comparable (Yokoi et al. 1982). Since some of the markers, such a T10, are also "activation" markers (Schlossman et al. 1983), the question of whether cord blood T cells actually have an immature versus an activated phenotype is still open. In the adult, 3% of T cells in human blood are T4+T8+ (Blue et al. 1985). However, coexpression of T4 and T8 is also related to activation, since the proportion of T4+T8+ cells rose to 60% after mitogen stimulation, was transitory, and the phenotype of these cells - although expressing other activation markers such as T9, T10, and IL-2 receptors - did not include T6, which is found in the majority of the T4+T8+ thymocytes (Blue et al. 1985). Whether this phenomenon is related to the immature peripheral T cells described above or to the progenitors described in the adult murine systems (Miller and Stutman 1984) needs further study. The possibility that peripheral T cells in the adult contain cells which may act as progenitors is attractive. The possibility of some T cells reverting to immature phenotypes as a consequence of activation (Blue et al. 1985) is intriguing and could offer a feed-forward type of regulation for peripheral T-cell renewal.

Although postthymic T cells are capable of extensive self-renewal, expansion, and function in the absence of the thymus, the self-renewal, expansion, and functional activity of prethymic cells in the absence of the thymus, although detectable, is very inefficient (Gorczynski and MacRae 1977; Galli and Droge 1980; Duprez et al. 1982; Piguet and Vassalli 1982; Miller and Stutman 1982, 1984). For example, prethymic cells generated only 2% of the pCTL when compared with postthymic T cells (Miller and Stutman 1984).

The same could be said about functional T cells in athymic nude mice (Maryanski et al. 1981; Chen et al. 1984). Extrathymic T-cell development in vivo in nude mice is detectable but very inefficient (assuming that indeed there is no temporary thymic exposure, Jenkinson et al. 1981). It takes months for nude mice to build up a small pool, 1%–2% of normal, of pCTL in the absence of a thymus (Maryanski et al. 1981; Chen et al. 1984). Conversely, during ontogeny a normal mouse would build up a much larger pool of precursor T cells in just 2–3 days in the presence of a thymus (Ceredig 1979; Ishizaka and Stutman 1983). Models of so-called in vitro differentiation of prethymic cells are not included in this discussion since they use T-cell products (IL-2 and other types of crude supernatants) and cannot be considered as true examples of extrathymic differentation (Stutman 1983).

In summary, postthymic T cells can express a thymus-independent state and can show extensive self-renewal, expansion, and functional activity in the absence of the thymus. While the degree of thymus dependency of posttyhmic expansion needs some study, it is apparent that "true" extrathymic T-cell differentiation in the absence of a thymus from putative prethymic cells is a highly inefficient procedure to generate functional T cells.

Table 2 shows that export of cells from the thymus is very active during the perinatal stages and at 30 days of age, but begins to decline at 4 months of age, with an almost total shut-off at 6 months of age, supporting studies by Scollay et al. (1980). The extensive thymus-independent postthymic development described above could be extremely important in the maintenance of the T-cell pool in periphery at times in which thymic export is negligible.

A low functional response can be due quantitative deficiencies of any or all of the cell components involved, which may or may not include T-cells, and to excess down regulation. Several problems intrinsic to these studies are worth mentioning. Maturation patterns differ between mouse strains at a given age and are affected by environmental conditions (Stutman 1985). Maturation patterns may even show regional variation at different organ sites at the same age and within the same strain (Ishizaka and Stutman 1983). For example, while the difference in frequency of IL-2-producing T cells between 6-day-old and adult mice was more than 20-fold in spleen it was only twofold in lymph nodes (Ishizaka and Stutman 1983). Individual variation was also evidenced by the broad scattering of results when human fetal lymphocytes were tested for their capacity to generate allo-specific CTL (Rayfield et al. 1980) or when mitogen responses in newborn infants were correlated with weight at birth (Toivanen et al. 1981). Another problem is the actual experimental design. For example, human fetal liver cells at 49 - 70 days of gestation showed strong proliferative responses to alloantigens in conventional mixed lymphocyte cultures (Stites et al. 1972, 1974). Such results would imply that alloreactivity could be detected in the human fetus before the development of the thymus. However, further studies showed that the response was due to back-stimulation by the adult T cells in the stimulator population (Toivanen et al. 1978).

There is agreement that all detectable T-cell functions appear in periphery approximately at birth or within the first perinatal weeks, in both mice and humans, and that activity is usually detected earlier in thymus (Stites et al. 1972, 1974; Toivanen et al. 1978, 1981; Stutman and Calkins 1977; Stutman 1985). In humans, once the responses are detected in periphery, their magnitude is usually comparable to the adult levels, while in mice the adult levels (usually tested in spleen) may be reached a few weeks later (Stutman and Calkins 1977). The actual detail of timings in the murine studies varies with strains and assays used and is probably not essential for our understanding of ontogeny (Stutman 1985). The immunological defects during early ontogeny, when detected, are usually multifactorial and related to both lower levels of precursors of the effector T cells and deficiencies of accessory cells (Stutman and Ishizaka 1982; Ishizaka and Stutman 1983). The possibility of qualitative differences in immunological capacity for T cells early during ontogeny has been already addressed. Finally, in spite of space limitations, the issue of suppressor cells during ontogeny has to be addressed for the sake of completeness, even if it seems to apply mostly to the murine spleen and human cord blood. I will not speculate on the teleological role of these regulatory cells, suffice it to say that in the embryo they may be involved in prevention of graft-versus-host reactions by the maternal cells, and that during the early perinatal period they may be involved in controlling possible deviations of the repertoire during periods of extensive cellular expansion (Stutman 1985).

5 Conclusion

The development of T cells during ontogeny follows some rules which are different from the mechanisms used by the adult for constant renewal of T cells in periphery.

Acknowledgments. The experimental work cited was primarily supported by NIH grants CA-08748, CA-17818 and AG-02152. This review was completed in September, 1985.

References

Abramson S, Miller RG, Phillips RA (1977) The identification in adult bone marrow of pluripotent and restricted stem cells of the myeloid and lymphoid systems. J Exp Med 145:1567-1579

Acuto O, Hussey RE, Fitzgerald KA, Protentis JP, Meuer SC, Schlossman SF, Reinherz EL (1983) The human T cell receptor: Appearance in ontogeny and biochemical relationship of a and b subunits on IL-2 dependent clones and T cell tumors. Cell 34:717-726

Augustin AA, Sim GK (1984) T-cell receptors generated via mutations are specific for various major histocompatibility antigens. Cell 39:5-12

Barclay AN, Mayrohofer G (1981) Bone marrow origin of Ia-positive cells in the medulla of rat thymus. J Exp Med 153:1660-1671

Battisto JR, Ponzio NM (1981) Autologous and syngeneic mixed lymphocyte reactions and their immunological significance. Progr Allergy 28:160-192

Basch RS, Berman JW (1982) Thy 1 determinants are present on many murine haemopoietic cells other than T cells. Eur J Immunol 12:359-364

Basch RS, Kadish JL, Goldstein G (1978) Hematopoietic thymocyte precursors. IV. Enrichment of the precursors and evidence for heterogeneity. J Exp Med 147:1843-1848

Bhan AK, Reinherz EL, Poppema S, McCluskey RT, Schlossman SF (1980) Location of T cell and major histocompatibility complex antigens in the human thymus. J Exp Med 152:771-782

Biggar WD, Stutman O, Good RA (1972) Morphological and functional studies of fetal thymus transplants in mice. J Exp Med 135:793-807

Blue ML, Daley JF, Levine H, Schlossman SF (1985) Coexpression of T4 and T8 on peripheral blood T cells demonstrated by two-color fluorescence flow cytometry. J Immunol 134:2281-2286

Born W, Yague J, Palmer E, Kappler J, Marrack P (1985) Rearrangement of T-cell receptor b-chain genes during T-cell development. Proc Natl Acad Sci USA 82:2925-2929

Bradley SM, Kruisbeek AM, Singer A (1982) Cytotoxic T lymphocyte responses in allogeneic bone marrow chimeras. The chimeric host strictly dictates the self-repertoire of Ia-restricted T cells but not H-2K/D-restricted T cells. J Exp Med 156:1650-1664

Cantor H, Boyse EA (1977) Regulation of Cellular and Humoral Immune Responses by T cell subclasses. Cold Spring Harbor Symp Quant Biol 41:23-32

Ceredig R (1979) Frequency of alloreactive cytotoxic T cell precursors in the mouse thymus and spleen during ontogeny. Transplantation 28:377-381

Ceredig R, Dialynas DP, Fitch FW, MacDonald HR (1983a) Precursors of T cell growth factor producing cells in the thymus: Ontogeny, frequency and quantitative recovery in a subpopulation of phenotypically mature thymocytes defined by monoclonal antibody GK-1.5. J Exp Med 158:1654-1671

Ceredig R, Lowenthal JW, Nabholz M, MacDonald HR (1985) Expression of interleukin-2 receptors as a differentiation marker of intrathymic stem cells. Nature 314:98-100

Ceredig R, MacDonald HR (1985) Intrathymic differentiation: Some unanswered questions. Surv Immunol Res 4:87-95

Ceredig R, MacDonald HR, Jenkinson EJ (1983b) Flow microfluorometric analysis of mouse thymus development in vivo and in vitro. Eur J Immunol 13:185-190

Ceredig R, Sekaly RP, MacDonald HR (1983c) Differentiation in vitro of Lyt 2+ thymocytes from embryonic Lyt 2- precursors. Nature 303:248-250

Chen WF, Scollay R, Clark-Lewis I, Shortman K (1983) The size of functional T-lymphocyte pools within thymic medullary and cortical cell subsets. Thymus 5:179-195

Chen WF, Scollay R, Shortman K, Skinner M, Marbrook J (1984) T cell development in the absence of the thymus: The number, the phenotype and the functional capacity of T lymphocytes in nude mice. Am J Anat 170:339-347

Ching LM, Miller TG (1981) Generation of cytotoxic T-lymphocyte precursor cells in T cell colonies grown in vitro. Nature 289:802-804

Ching LM, Miller RG (1982) Development of cytotoxic T lymphocyte precursor cells in T cell colonies grown in vitro. J Immunol 129:2345-2351

Davis MM, Chien YH, Gascoigne NRJ, Hedrick SM (1984) A murine T cell receptor gene complex: Isolation, structure and rearrangement. Immunol Rev 81:235-258

Decker JM, Clarke J, Bradley LM, Miller A, Secrarz EE (1974) Presence of antigen-binding cells for five diverse antigens at the onset of lymphoid development: lack of evidence for somatic diversification during ontogeny. J Immunol 113:1823-1833

Doherty PC, Korngold R, Schwartz DH, Bennink JR (1981) The development and loss of virus-specific thymic competence in bone marrow chimeras and normal mice. Immunol Rev 58:37-72

Duprez V, Hamilton B, Burakoff SJ (1982) Generation of cytolytic T lymphocytes in thymectomized, irradiated and bone marrow-reconstituted mice. J Exp Med 156:844-859

Dwyer JM, MacKay IR (1972a) The development of antigen binding lymphocytes in foetal tissues. Immunology 25:871-879

Dwyer JM, MacKay IR (1972b) Validation of autoradiography for recognition of antigen-binding lymphocytes in blood and lymphoid tissues. Clin Exp Immunol 10:581-597

Everett NB, Caffrey RW, Rieke WO (1964) Recirculation of lymphocytes. Ann NY Acad Sci 113:887-897

Ezine S, Weissman IL, Rouse RV (1984) Bone marrow cells give rise to distinct cell clones within the thymus. Nature 309:629-631

Fink PJ, Gallatin MW, Reichert RA, Butcher EC, Weissman IL (1985) Homing receptor-bearing thymocytes: An immunocompetent cortical subpopulation. Nature 313:233-235

Fontaine-Perus JC, Calman FM, Kaplan C, LeDouarin NM (1981) Seeding of the 10-day mouse embryo thymic rudiment by lymphocyte precursors in vitro. J Immunol 126:2310-2316

Ford CE (1966) Traffic of lymphoid cells in the body. In: The thymus: experimental and clinical studies. Ciba Found Symposium. Little, Brown, Boston, p 131

Fowlkes BJ, Mathieson BJ (1985) Intrathymic differentiation: Thymocyte heterogeneity and the characterization of early T-cell precursors. Surv immunol Res 4:96-109

Galli P, Droge W (1980) Development of cytotoxic T lymphocyte precursors in the absence of the thymus. Eur J Immunol 10:87-92

Gorczynski RM, MacRae S (1977) Differentiation of functionally active mouse T lymphocytes from functionally inactive bone marrow precursors. Immunology 33:697-712

Greiner DL, Goldschneider I, Barton RW (1982) Identification of thymocyte progenitors in hemopoietic tissues of the rat. II. Enrichment of functional prothymocytes on the fluorescence-activated cell sorter. J Exp Med 156:1448-1460

Griffiths-Chu S, Patterson JAK, Berger CL,Edelson Rc, Chu AC (1984) Characterization of immature T cell subpopulations in neonatal blood. Blood 64:296-300

Haynes BF (1984) The human thymic microenvironment. Adv Immunol 36:87-142

Howe ML, Goldstein AL, Battisto J (1970) Isogenic lymphocyte interaction: recognition of self antigens by cells of the neonatal thymus. Proc Natl Acad Sci USA 67:613-619

Huber B, Cantor H, Shen FW, Boyse EA (1976) Independent differentiative pathways of Ly1 and Ly23 subclasses of T cells. Experimental production of mice deprived of selected T cell subclasses. J Exp Med 144:1128-1132

Ishizaka ST, Stutman O (1983) Analysis by limiting dilution of interleukin 2-producing T cells in murine ontogeny. Eur J Immunol 13:936-942

Janossy G (1983) Human lymphocyte subsets: their tissue distribution. In: Yamamura Y, Tada T (eds) Progress in immunology V. Academic, Tokyo, p 1085

Jenkinson EJ, VanEwijk W, Owen JJT (1981) Major histocompatibility complex antigen expression on the epithelium of the developing thymus in normal and nude mice. J Exp Med 153:280-292

Joel DD, Chanana AD, Cronkite EP (1974) Thymus cell migration. Ser Haematol 7:464–481

Jotereau FV, LeDouarin NM (1982) Demonstration of a cyclic renewal of the lymphocyte precursor cells in the quail thymus during embryonic and perinatal life. J Immunol 129:1869–1877

Kadish JL, Basch RS (1976) Hematopoietic thymocyte precursors. I. Assay and kinetics of the appearance of progeny. J Exp Med 143:1082–1099

Kamarck ME, Gottlieb PD (1977) Expression of thymocyte surface alloantigens in the fetal mouse thymus in vivo and in organ culture. J Immunol 119:407–415

Kendall MD (1981) The cells of the thymus. In: Kendall MD (ed) The thymus gland. Academic, New York, p 63

Komuro K, Boyse EA (1973) Induction of T lymphocytes from precursor cells in vitro by a product of the thymus. J Exp Med 138:479–482

Komuro K, Goldstein G, Boyse EA (1975) Thymus-repopulating capacity of cells that can be induced to differentiate to T cells in vitro. J Immunol 115:195–198

Kruisbeek AM, Sharrow SO, Singer A (1983) Differences in the MHC-restricted self-recognition repertoire of intra-thymic and extra-thymic cytotoxic T lymphocyte precursors. J Immunol 130:1027–1032

Lajtha LG (1966) Cytokinetics and regulation of progenitor cells. J Cell Physiol 67, Suppl. 1:133–148

Lattime EC, Gillis S, David C, Stutman O (1981) Interleukin 2 production in the syngeneic mixed lymphocyte reaction. Eur J Immunol 11:67–69

Lattime EC, Gillis S, Pecoraro G, Stutman O (1982) Ia-dependent Interleukin 2 production in syngeneic cellular interactions. J Immunol 128:480–485

Lattime EC, Golub SH, Stutman O (1980) Lyt 1 cells respond to Ia bearing macrophages in the murine syngeneic mixed lymphocyte reaction. Eur J Immunol 10:723–726

Lattime EC, Macphail S, von Wussow P, Stutman O (1983) Ia dependent lymphokine production in the syngeneic mixed lymphocyte response. Behring Inst Mitt 72:47–58

LeDouarin NM (1977) Ontogeny of primary lymphoid organs. In: Loor F, Roelants GE (eds) B and T cells in immune recognition. Wiley, New York, p 1

Lepault F, Coffman RL, Weissman IL (1983) Characteristics of thymus-homing bone marrow cells. J Immunol 131:64–69

Longo DL, Schwartz RH (1980) T cell specificity for H-2 and Ir phenotype correlates with the phenotype of thymic antigen-presenting cells. Nature 287:44–46

Maccario R, Ferrari FA, Siena S, Vitiello MA, Martini A, Siccardi AG, Ugazio AG (1981) Receptors for peanut agglutinin on a high percentage of human cord-blood lymphocytes: phenotype characterization of peanut-positive cells. Thymus 2:239–337

Maccario R, Nespoli L, Mingrat G, Vitiello A, Ugazio AG, Burgio GR (1983) Lymphocyte subpopulations in the neonate: identification of an immature subset of OKT8-positive, OKT3-negative cells. J Immunol 130:1129–1131

Mandel T, Russell PJ (1971) Differentiation of fetal mouse thymus: ultrastructure of organ cultures and subcapsular grafts. Immunology 21:659–674

Martin C, Beaupain D, Dieterlen-Lievre F (1978) Developmental relationships between viteline and intra-embryonic haemopoiesis studied in avian "yolk sac chimeras". Cell Differentiation 7:115–130

Maryanski JL, MacDonald HR, Sordat B, Cerottini JC (1981) Cytolytic T lymphocyte precursor cells in congenitally athymic C57BL/6 nu/nu mice: Quantitation, enrichment and specificity. J Immunol 129:521–525

Mathieson BJ, Fowlkes BJ (1984) Cell surface antigen expression on thymocytes: development and phenotypic differentiation of intrathymic subsets. Immunol Rev 82:141–143

Mathieson BJ, Sharrow SO, Rosenberg Y, Hammerling U (1981) Lyt 1+23– cells appear in the thymus before Lyt 1, 2, 3+ cells. Nature 289:179–181

Metcalf D (1977) Haemopoietic colonies. In vitro cloning of normal and leukemic cells. Springer, New York Heidelberg Berlin

Micklem HS, Ford CE, Evans EP, Gray J (1966) Interrelationships of myeloid and lymphoid cells: Studies with chromosome-marked cells transfused into lethally irradiated mice. Proc R Soc 165:78–102

Micklem HS, Ford CE, Evans EP, Ogden DA, Papworth DS (1972) Competitive in vivo proliferation of foetal and adult haematopoietic cells in lethally irradiated mice. J Cell Physiol 79:293–298

Miller III JJ (1969) Studies of the phylogeny and ontogeny of the specialized lymphatic tissue venules. Lab Invest 21:484–490

Miller RA, Stutman O (1982) Generation of functionally distinct T cell subsets from post-thymic T cell progenitors in vivo. Behring Inst Mitt 70:101–105
Miller RA, Stutman O (1984) T cell repopulation from functionally restricted splenic progenitors: 10000-fold expansion documented by limiting dilution analyses. J Immunol 133:2925–2932
Möller G (ed) (1978) Acquisition of the T cell repertoire. Immunol Rev 42
Moore MAS, Owen JJT (1967) Experimental studies on the development of the thymus. J Exp Med 126:715–726
Morrisey PJ, Kruisbeek AM, Sharrow SO, Singer A (1982) Tolerance of thymic cytotoxic T lymphocytes to allogeneic H2 determinants encountered prethymically: Evidence for expression of anti-H2 receptors prior to entry into the thymus. Proc Natl Acad Sci USA 79:2003–2007
Owen JJT, Jenkinson EJ (1981) Embryology of the Lymphoid System. Prog Allergy 29:1–34
Piguet PF, Vassalli P (1982) Long term reconstituion of "B"mice with post-thymic T cells: Expansion of the post-thymic T cells and absence of extra thymic maturation. Behring Inst Mitt 70:96–100
Piguet PF, Irle C, Kollate E, Vassalli P (1981) Post-thymic T lymphocyte maturation during ontogenesis. J Exp Med 154:581–593
Raulet DH (1985) Expression and function of interleukin-2 receptors on immature thymocytes. Nature 314:101–103
Raulet DH, Garman RD, Saito H, Tonegawa S (1985) Developmental regulation of T-cell receptor gene expression. Nature 314:103–107
Rayfield LS, Brent L, Rodeck CH (1980) Development of cell-mediated lympholysis in human foetal blood lymphocytes. Clin Exp Immunol 42:561–570
Reichert RA, Gallatin WM, Butcher EC, Weissman IL (1984) A homing receptor- bearing cortical thymocyte subset: implications for thymus cell migration and the nature of cortisone-resistant thymocytes. Cell 38:89–99
Reinherz EL, Schlossman SF (1980) The differentiation and function of human T lymphocytes. Cell 19:821–827
Reinherz EL, Kung PC, Goldstein G, Levey RH, Schlossman SF (1980) Discrete stages of human intrathymic differentiation: analysis of normal thymocytes and leukemic lymphoblasts of T-cell lineage. Proc Natl Acad Sci USA 77:1588–1592
Reinherz EL, Meuer S, Fitzgerald KA, Hussey RE, Levine H, Schlossman SF (1982) Antigen recognition by human T lymphocytes is linked to surface expression of the T3 molecular complex. Cell 30:735–743
Rocha BA, Freitas A, Coutinho A (1983) Population dynamics of T lymphocytes. Renewal rate and expansion in the peripheral lymphoid organs. J Immunol 131:2158–2164
Roehm N, Herron L, Cambier J, DiGuisto D, Haskins K, Kappler J, Marrack P (1984) The major histocompatibility complex-restricted antigen receptor on T cells: distribution on thymus and peripheral cells. Cell 38:577–584
Rouse RV, Van Ewijk W, Jones PP, Weissman IL (1979) Expression of MHC antigens by mouse dendritic cells. J Immunol 122:2508–2515
Samelson LE, Lindsten T, Fowlkes BJ, van den Elsen P, Terhorst C, Davis MM, Germain RN, Schwartz RH (1985) Expression of genes of the T-cell antigen receptor complex in precursor thymocytes. Nature 315:765–768
Scheid MP, Hoffmann MK, Komuro K, Hammerling U, Abbott J, Boyse EA, Cohen G, Hooper JA, Schulof RA, Goldstein AL (1973) Differentiation of T cells induced by preparations from thymus and by nonthymic agents. J Exp Med 138:1027–1032
Scheid MP, Goldstein G, Boyse EA (1978) The generation and regulation of lymphocyte populations. Evidence from differentiative induction systems in vitro. J Exp Med 147:1727–1743
Schlossman SF, Meuer SC, Acuto O, Morimoto C, Reinherz EL (1983) Human T lymphocytes: their differentiative history and functional program. In: Yamamura Y, Tada T (eds) Progress in immunology V. Academic, Tokyo, p 1069
Scollay R (1982) Thymus cell migration: cells migrating from thymus to peripheral lymphoid organs have a "mature" phenotype. J Immunol 128:1566–1570
Scollay R, Shortman K (1983) Thymocyte subpopulations: an experimental review, including flow cytometric cross-correlations between the major murine thymocyte markers. Thymus 5:245–295
Scollay T, Shortman K (1984) Contribution of thymus lymphocytes to the peripheral lymphoid tissues and the effect of antigen on the rate of cell exit from the thymus. Am J Anat 170:331–338

Scollay T, Shortman K (1985) Identification of early stages of T lymphocyte development in the thymus cortex and medulla. J Immunol 134:3632–3642
Scollay R, Bartlett P, Shortman K (1984a) T cell development in the adult murine thymus: Changes in the expression of the surface antigens Ly2, L3t4 and B2A2 during development from early precursor cells to emigrants. Immunol Rev 82:79–103
Scollay R, Butcher EC, Weissman IL (1980) Thymus cell migration: Quantitative aspects of cellular traffic from the thymus to the periphery in mice. Eur J Immunol 10:210–218
Scollay R, Chen WF, Shortman K (1984b) The functional capabilities of cells leaving the thymus. J Immunol 132:25–30
Scollay R, Wilson A, Shortman K (1984c) Thymus cell migration: Analysis of thymus emigrants with markers that distinguish medullary thymocytes from peripheral T cells. J Immunol 132:1089–1094
Singer A, Bradley SM, Kruisbeek AM (1982) Role of the thymus in dictating the self-repertoire of T cells which differentiate within it: Evidence for both an intra-thymic and an extra-thymic differentiation. Pathway for T cells restricted to KD, but not for T cells restricted to I. Behring Inst Mitt 70:140–146
Shortman K, Mandel T, Andrews P, Scollay R (1985) Are any functionally mature cells of medullary phenotype located in the thymus cortex? Cell Immunol 933:350–363
Smith K (1984) Inside the thymus. Immunol Today 5:83–84
Snodgrass HR, Kisielow P, Kiefer M, Steinmetz M, von Boehmer H (1985) Ontogeny of T-cell antigen receptor within the thymus. Nature 313:592–595
Solomon JB (1978) Immunological milestones in ontogeny. Develop Compar Immunol 2:409–424
Stites DP, Carr M, Fudenberg HH (1974) Ontogeny of cellular immunity in the human fetus. Development of responses to phytohemagglutinin and to allogeneic cells. Cell Immunol 11:257–271
Stites DP, Wybran J, Carr MC, Fudenberg HH (1972) Development of cellular immune competence in man. In: Ontogeny of acquired immunity. A Ciba Foundation Symposium. Elsevier-Excerpta Medica-North Holland, Amsterdam p 113
Stutman O (1972) Traffic of cells and development of immunity. In: Day SB, Good RA (eds) Membranes and virus in immunopathology. Academic, New York p 437
Stutman O (1975) Humoral thymic factors influencing postthymic cells. Ann NY Acad Sci 249:89–104
Stutman O (1976) Migration of yolk sac cells to thymus grafts: Requirement of prior sojourn in bone marrow (or liver?). Ann Immunol (Inst Pasteur) 127C:943–949
Stutman O (1977) Two main features of T-cell development: Thymus traffic and postthymic maturation. Contemp Topics Immunobiol 7:1–46
Stutman O (1978) Intrathymic and extrathymic T cell maturation. Immunol Rev 42:138–184
Stutman O (1983) Role of thymic hormones in T cell differentiation. Clin Immunol Allergy 3:9–91
Stutman O (1985) Ontogeny of T Cells. Clin Immunol Allergy 5:191–234
Stutman O, Calkins CE (1977) Ontogenetic aspects (of transplantation). In: Masshoff JW (ed) Handbuch der allgemeinen Pathologie VI/8. Transplantation. Springer, Berlin Heidelberg New York, p 163
Stutman O, Carnaud, Ishizaka ST, Lattime EC, Miller RA (1982) Intra and extrathymic differentiation of T cells. In: Serrou B, Rosenfeld C, Daniels JC, Saunders JP (eds) Current concepts in human immunology and cancer immunomodulation. Elsevier, Amsterdam p 3
Stutman O, Good RA (1971) Immunocompetence of embryonic hemopoietic cells after traffic to thymus. Transplant Proc 3:923–925
Stutman O, Ishizaka ST (1982) Ontogeny of T cell function: alloreactivity appears earlier than reactivity against hapten-modified self and interleukin-2 production. Clin Immunol Immunopathol 23:202–214
Stutman O, Shen FW (1979) Postthymic precursor cells give rise to both Lyt-1 and Lyt-2,3 subsets of T cells. Transplant Proc 11:907–909
Stutman O, Lattime EC, Gillis S, Miller RA (1980) Role of interleukin 2 in postthymic maturation of T cells. Behring Inst Mitt 67:95–104
Toivanen P, Asantila T, Granberg C, Leino A, Hirvonen T (1978) Development of T cell repertoire in the human and the sheep fetus. Immunol Rev 42:185–201
Toivanen P, Uksila J, Leino A, Lassila O, Hirvonen T, Ruuskanen O (1981) Development of mitogen responding T cells and natural killer cells in the human fetus. Immunol Rev 57:89–105
Trowbridge IS, Lesley J, Trotter J, Hyman R (1985) Thymocyte subpopulation enriched for progenitors with an unarranged T-cell receptor beta-chain gene. Nature 315:666–669

van den Elsen P, Shepley BA, Cho M, Terhorst C (1985) Isolation and characterization of a cDNA clone encoding the murine homologue of the human 20K T3/T-cell receptor glycoprotein. Nature 314:542–544

Van Ewijk W, Jenkinson EJ, Owen JJT (1982) Detection of Thy 1, T-200, Lyt 1 and Lyt 2-bearing cells in the developing lymphoid organs of the mouse embryo in vivo and in vitro. Eur J Immunol 12:262–271

Van Ewijk W, Rouse RV, Weissman IL (1980) Distribution of H-2 microenvironments in the mouse thymus. Immunoelectron microscopic identification of I-A and H-2K bearing cells. J Histochem Cytochem 28:1089–1099

Wagner H, Rollinghoff M, Pfizenmaier K (eds) (1982) The Influence of the Thymus on the Generation of the T Cell Repertoire. Behring Inst Mitt 70

Wallis VJ, Leuchars E, Chwalinsky S, Davies AJS (1975) On the sparse seeding of bone marrow and thymus in radiation chimeras. Transplantation 19:2–11

Wilson M (1985) Immunology of the fetus and the newborn: lymphocyte phenotype and function. Clin Immunol Allergy 5:271–286

Yokoi T, Miyawaki T, Yachie A, Ohzeki S, Taniguchi N (1982) Discrepancy in expression ability of Tac antigen and Ia determinants defined by monoclonal antibodies on activated or cultured cord blood T lymphocytes. J Immunol 129:1441–1445

Yoshikai Y, Yanagi Y, Suciu-Foca N, Mak TW (1984) Presence of T-cell receptor mRNA in functionally distinct T cells and elevation during intrathymic differentiation. Nature 310:506–508

Zinkernagel RM (1978) Thymus and lymphohemopoietic cells: Their role in T cell maturation and selection of T cells' H-2-restriction-specificity and in H-2 linked Ir gene control. Immunol Rev 42:224–270

Humoral Immunity in the Human Neonate

JOSEPH L. BUTLER, TAKAYO SUZUKI, HIROMI KUBAGAWA, and MAX D. COOPER

Cellular Immunobiology Unit of the Tumor Institute, Departments of Pediatrics, Microbiology and Pathology, and The Comprehensive Cancer Center, University of Alabama at Birmingham, Birmingham, AL 35294, USA

1 Introduction

The human neonate is highly susceptible to invasion by microorganisms and as a result has been the focus of intense immunological investigation. Ontogenic studies have delineated the origin and development of immunocompetent B cells (Cooper et al. 1966; Osmond and Nossal 1974; Kamps and Cooper 1982), yet the precise definition of B cell function in early postnatal life remains incomplete. As a result of clinical and experimental observations, several distinctive features of the neonatal immune system have been described. Newborns fail to produce normal amounts of immunoglobulin to polysaccharide antigens, and when they are immunized with protein antigens, their antibody responses are primarily restricted to the IgM isotype. There is a relatively slow progression from IgM to IgG production, and IgA responses are the last to mature. This developmental lag is reflected in the sequential attainment of adult levels of serum immunoglobulins: IgM, 1 year; IgG, 5–6 years; and IgA, 10–14 years. The delay in antibody production is not due to lack of precursor B cells, since the newborn has large numbers of circulating B cells bearing surface IgM, IgG, and IgA. Thus, a gap appears to exist between the development of lymphocytes and their ability to cooperate effectively in generating normal antibody responses.

Any consideration of the newborn immune system is dependent on an understanding of the origin, development, and function of immunocompetent cells. In this chapter, the pathways of normal B- and T-cell development will be reviewed, and the factors which may contribute to impaired humoral immunity in the neonate will be addressed.

2 B Cell Development

The development of B-lineage cells can be considered in two distinct phases: an antigen-independent pre-B phase and an antigen-dependent B phase. Pluripotential stem cells of mesenchymal origin initially give rise to erythroid and myeloid cells in blood islands of the yolk sac and later migrate to the fetal liver where differentiation along B-cell lineage begins (Gathings et al. 1977; Kamps and Cooper 1982). Bone marrow assumes B-cell hematopoiesis later in fetal development and continues this function throughout life.

Immunology of the Neonate
Burgio, Hanson, Ugazio (Eds.)

Differentiation of B cells begins when pluripotential stem cells give rise to large, immature pre-B cells which do not express immunoglobulin heavy or light chains. Large pre-B cells undergo an organized sequence of DNA rearrangements of the immunoglobulin variable-region genes, and as these rearrangements occur the cells proliferate in a polyclonal fashion (Rossant et al. in press). Heavy-chain variable-region genes on chromosome 14 are selected for the first DNA rearrangements by translocation of a diversity (D) gene in juxtaposition with a J_H gene and deletion of the interposed DNA (Tonegawa 1983; Alt et al. 1983). Next, the complete variable region of the heavy chains is encoded by translocating a V_H to the DJ_H segment to form a VDJ complex. Such an arrangement allows putative transcriptional enhancer sequences to influence the promoter sequence 5' of the transposed V_H leader sequence (Gillies et al. 1983; Wabl and Burrows 1984, Zaller and Eckhardt 1985). The complete $C\mu$ gene complex is then transcribed and the transcript processed into μ-chain mRNA. When the VDJ_H rearrangements are complete, a signal is given to begin rearrangement of light-chain variable-region genes. These rearrangements occur primarily in small, postmitotic pre-B cells and are usually initiated in the kappa gene loci on chromosome 2 (Hieter et al. 1981). Transcription begins when a productive VJ_L rearrangement occurs in one light-chain gene locus. When light-chain mRNA is formed from the transcript, further V-J_L rearrangements on other chromosomes are halted. If a productive VJ_K rearrangement fails on both chromosomes 2, the rearrangement mechanism proceeds to the lambda family on chromosome 22. Normal individuals generate millions of pre-B-cell clones during this phase of polyclonal expansion. Each clone differs in its immunoglobulin heavy- and light-chain genes and in the antibody specificity that these genes determine.

The antigen-dependent B-cell phase is heralded by light-chain synthesis and expression of surface-membrane-bound IgM molecules. The expression of surface immunoglobulin molecules is a unique feature of B cells and is necessary for their activation and clonal selection by antigen. Newly formed B cells leave the bone marrow microenvironment and migrate to the spleen, lymph nodes, and other lymphoid tissues where a substantial proportion die. Other B cells grow and differentiate in a pauci-clonal fashion under the influence of antigen-presenting cells and T cells. Surface receptors for the Fc portion of IgG molecules, Epstein-Barr virus (EBV) and complement components (C3b and C3d) are expressed as B cells differentiate. In addition, HLA-DR molecules are expressed on both B and pre-B cells. These surface glycoproteins are genetically polymorphic recognition molecules that facilitate T-B-cell interaction and subsequent B-cell activation.

Some members of each B-cell clone undergo a switch in the isotype of the immunoglobulin produced, changing from IgM and IgD to IgG, IgA or IgE expression. It is during this last step in immunoglobulin gene rearrangement that the switch (S) region 5' of $C\mu$ recombines with an S region 5' of a downstream C_H gene (Davis et al. 1980; Honjo 1983). This recombination event results in the coexpression of surface IgM and non-IgM isotypes and appears to be independent of antigenic stimulation or T-cell help (Abney et al. 1978; Bankhurst and Warner 1972). On the other hand, terminal plasma cell differentiation and the accompanying transition from membrane expression to secretion of different (non-IgM) heavy chain isotypes is influenced by antigens, T cells, accessory cells, and soluble factors (Forni and Coutinho 1982). The role of T cells in isotype switching remains unclear. There is evidence from studies by Mayumi et al. (1983) and Kiyono et a. (1984) for preferential T-cell help to isotype-committed B-cells. In contrast, Mayer et al. (1985) have reported the existence of T cells that are able to induce isotype switching.

Following activation by antigens or mitogen, B cells undergo predictable changes. Activated B cells enlarge, proliferate, and express receptors for soluble growth and differentiation factors elaborated by T cells (Kishimoto 1985). As proliferation and terminal differentiation proceed to the mature plasma cell stage, surface immunoglobulin, HLA-DR molecule and complement receptor expression decline as high rate synthesis and secretion of immunoglobulin molecules ($>10^3$/sec) begin. Plasma cells rarely divide and they have short life spans of a few days; their B-cell precursors must be continually replaced by hematopoiesis in the bone marrow. A subpopulation of activated B-cell clones do not complete terminal plasma cell differentiation. These cells, termed memory B cells, are characterized by their long life span and heightened antibody responses on secondary exposure to antigens.

3 T Cell Development

A discussion of humoral immunity would be incomplete without a consideration of T-cell participation. Therefore, T-cell development and functional T–B cell interactions will be briefly described. T cells derive their name from their site of differentiation in the thymus and arise from the pluripotential stem cell of hemopoietic tissue. The thymus develops early in embryonic life from endodermal cells of the third and fourth pharyngeal pouches (Norris 1938). During the 8th week of gestation precursor stem cells, lacking T-cell antigens, leave the circulation to enter the thymic microenvironment (LeDouarin and Jotereau 1975). There, they are stimulated to proliferate under the influence of thymic hormones or through direct contact with the thymic epithelium. Rapidly dividing, immature thymocytes enter the medulla and cease to proliferate. The migration from cortex to medulla is characterized by changing expression of surface-bound glycoproteins (Reinherz et al. 1980). These surface molecules can be easily detected with monoclonal antibodies and serve as useful markers of T-cell differentiation. A small subpopulation of cortical thymocytes express transferrin receptors on their surface. As differentiation proceeds in the medullary regions, transferrin receptor expression is replaced by the T_3, T_4, T_6 and T_8 surface antigens. The T_6 molecule characterizes intrathymic T cell differentiation and is not expressed by circulating T cells. In contrast, the T_3 antigen is acquired relatively late in the differentiation process. This 20 KD glycoprotein is found on only 40% of the most mature thymocytes but is expressed by all mature T cells. Expression of the T_4 and T_8 surface antigens changes as differentiation proceeds. These antigens are coexpressed on approximately 70% of thymocytes; however, dual expression is absent on mature T cells. Thus, approximately 70% of peripheral blood T cells have the helper or inducer T_4 marker with the remainder expressing the suppressor or cytotoxic T_8 phenotype. Phenotype does not absolutely correlate with function as suppressor T cells may be T_4 positive, and cells expressing the T_4 antigen may have cytotoxic properties (Thomas et al. 1981; Moretta et a. 1981).

Mature T cells leave the thymus and travel throughout the body via the circulation, and subpopulations of these cells enter peripheral lymphoid organs such as the spleen, lymph nodes, or Peyer's patches of the intestine. The mechanisms of preferential attraction of T-cell subpopulations to peripheral lymphoid organs remains unclear; however, T-cell homing may be facilitated by the interaction between determinants on endothelial cells of lymphatic channels and receptors on the surface of T cells. The T cells remain in peripheral lymphoid tissues for variable periods and eventually reenter the circulation via lymphatic channels and their venous connections.

Antigens presented on the surface of accessory cells are recognized by specific T-cell receptors. The genes encoding these receptors have been studied extensively, and regions with structural similarity to the V, J, and C regions of immunoglobulin genes have been identified. Thus far, three T-cell-specific cDNA clones have been described: α, β, and γ. The α and β clones encode components of the T-cell antigen receptor while the γ polypeptide appears to be unrelated to the structure of this receptor. Rearrangements of genomic DNA occur during T cell differentiation and have been best studied in the DNA encoding the β polypeptide (reviewed by Hood et al. 1985). The generation of the $V\beta$ gene occurs as a result of rearrangement of $V\beta$, $D\beta$, and $J\beta$ gene segments in a fashion analogous to immunoglobulin gene rearrangements. Available evidence suggests that similar mechanisms are operative in the generation of $V\alpha$ genes. Transcription of the γ gene occurs primarily in cytotoxic T cells suggesting that the γ polypeptide may comprise part of a receptor complex for class I MHC molecules.

4 Fetal Immunity

Stimulation with exogenous antigen probably occurs rarely in the developing human, and consequently the functional status of the normal fetal immune system has not been completely defined. Previous studies have demonstrated that immunoglobulin synthesis begins around 20 weeks gestation (Alford et al. 1974). Cell-mediated immunity matures prior to the end of gestation, as evidenced by the capacity of newborns to reject foreign skin grafts (Fowler et al. 1960). Similarly, surveillance against foreign lymphocytes is operative during fetal life, as intrauterine blood transfusions rarely elicit graft-versus-host disease in the fetal recipient (Parkman et al. 1974).

Host responses to congenital infections have provided the best evidence for immunological maturity of the fetus. Infected fetuses make specific antibodies to invading microorganisms, with IgM being the predominant isotype. Since IgM does not cross the placenta, the contribution of maternal antibody to this IgM response is probably nil in the absence of a placental leak. Fetal IgA production may be increased by congenital infections, particularly cytomegalovirus (Mills et al. 1981). The contribution of fetal IgG in the response to infection is difficult to ascertain as the fetal antibodies are overshadowed by transplacentally derived maternal IgG.

5 Antibody Responses in the Newborn

Humoral immunity in the neonate appears deficient on the basis of: (a) increased incidence of bacterial infections; (b) poor antibody response to immunization with polysaccharide antigens, and (c) limited isotype response to protein antigens. Improved definition of this relative immunodeficiency has been provided by studies of neonatal B-cell differentiation. Several studies have focused on differentiation responses of cord blood lymphocytes (CBL) to PWM. This T-dependent polyclonal B-cell mitogen can induce differentiation of plasma cells of all isotypes. PWM stimulates CBL to proliferate normally yet a delay occurs in B-cell differentiation resulting in decreased production of plasma cells (Hayward and Lawton 1977), markedly diminished secretion of antibody, and restriction of secreted antibody to the IgM isotype (Morito et al. 1979). This finding may be the result of low frequencies of responsive B-cells in the newborn rather than inadequate T-cell help (Wu et al. 1976). However, Mayer (personal communication) has been able to trigger the production of IgM, IgG, and IgA by purified newborn B-cells in the presence of adult T-cells, suggesting that T-cell help is deficient in the newborn.

Neonatal B-cell differentiation has also been studied using T-cell-independent B cell mitogens. Epstein-Barr virus is a herpesvirus that has been extensively studied. It infects one out of three peripheral blood B cells exposed in vitro and stimulates immunoglobulin synthesis in approximately 10% of those infected cells. When EBV is used to stimulate cord B cells, IgM production in the absence of demonstrable IgG or IgA has been widely reported (Steel et al. 1977; Andersson et al. 1980; Tosato et al. 1980). In contrast, Konowalchuk et al. (1982) found that EBV-stimulated neonatal B cells secreted IgM and IgG at levels comparable to transformed adult B cells. *Staphylococcus aureus* Cowan-I (SAC) also bypasses the requirement for T-cell help in the activation of resting B cells by directly crosslinking surface Ig (Romagnani et al. 1981). Using SAC stimulation, Ruuskanen et al. (1980) reported secretion of IgM and IgA, but not IgG, in cultures of cord B cells. In addition, the amount of secreted antibody was comparable to that produced by cultures of adult B cells.

Andersson et al. (1981) have investigated the maturation of IgG subclass responses using PWM and EBV as promoters of B cell differentiation. PWM stimulates adult cells to produce IgG_1 with relatively little IgG_2, IgG_3, and IgG_4. On the other hand, EBV induces a response pattern in the order of $IgG_3 > IgG_1 > IgG_4 > IgG_2$. The capacity to synthesize the various IgG subclasses varies with age. Cord B cell cultures stimulated with EBV produced cytoplasmic IgG_1 and IgG_3. However, no IgG_2 or IgG_4 was demonstrated, and synthesis of these antibodies remained deficient form birth to at least 24 months of age. The neonate's poor response to polysaccharide antigens may be related to the delayed acquisition of IgG_2, as this IgG subclass appears to be the predominant antibody produced to carbohydrate antigens.

The mechanisms accounting for the failure of neonatal B cells to differentiate normally are not well understood and could be operative at various phases: early B cell activation, clonal expansion of activated B cells, or terminal differentiation. In recent years, the role of signals provided to B cells by surface Ig crosslinkage and by soluble protein mediators produced by activated T cells has been intensively investigated (Kishimoto 1985). Lymphokines, termed B-cell growth factors (BCGF) and B-cell differentiation factors (BCDF), have been well characterized and are important in the orderly regulation of B-cell function. Development of monoclonal antibodies to growth and differentiation factor receptors on the surface of B cells provides an excellent means of analyzing selected phases of normal and aberrant differentiation. In this regard, we have produced a mouse monoclonal antibody, 9.4, that reacts with an antigenic determinant of a surface molecule of 59 KD molecular weight (Suzuki et al. submittet for publication). The 9.4 antigen appears to be unique to activated, mature B cells and is absent on other B lineage cells. The expression of this surface antigen is induced by B-cell activation with a variety of polyclonal activators such as SAC, anti-μ and phorbol esters.

In vivo preactivated B cells expressing the 9.4 antigen have been demonstrated in peripheral blood (normal and leukemic), tonsils, and spleen. Evidence suggests that the antigen recognized by the 9.4 antibody is a component of the receptor for BCGF-I (low-molecular-weight BCGF). The 9.4 antibody inhibits the expected synergistic proliferation of B cells activated with anti-μ and cultured with a monoclonal T cell hybridoma-derived BCGF (Butler et al. 1983) without affecting anti-μ-induced B-cell activation. It is possible that abnormalities in growth factor receptor expression could play a role in deficient differentiation responses of neonatal B cells. Thus, we are currently investigating the expression of BCGF receptors on newborn B cells using the 9.4 antibody as well as the responsiveness of these cells to polyclonal activators and BCGF.

6 T Cell Regulation of Neonatal B Cells

As early as 12 weeks' gestation, T cells can be activated by antigens presented on the surface of antigen-processing cells, such as macrophages, dendritic cells, and Langerhans cells, and by soluble products of these accessory cells (Hayward and Lawton 1977). In turn, activated T cells secrete a variety of low-molecular-weight protein mediators whose functions include the stimulation of T-cell growth; enhancement of B-cell proliferation and differentiation into antigen-secreting plasma cells; and recruitment of macrophages, natural killer cells, and neutrophils to participate in the immune response. Alternatively, suppressor T cells exert negative control over these host defense mechanisms via secretion of soluble suppressor substances or by direct inhibitory cell-to-cell contact. Thus, the T cell assumes primary responsibility for the complex mechanisms involved in normal immunoregulation.

Abnormalities in cellular immunity may contribute to the deficient antibody response in the neonate. Previous experiments by Hayward and Kurnick (1981) report that active T suppressor cells bearing surface Fc receptors for IgM occur in cord blood as early as 26 weeks of gestation and are capable of inhibiting proliferation and Ig secretion by maternal lymphocytes (Oldstone and Tishon 1977). These suppressor T cells are found in increased frequency in peripheral blood of neonates and older infants and gradually decrease to near adult levels by 3 months of age (Miyawaki et al. 1979). Mixing experiments demonstrate that normal adult B cells fail to differentiate in response to PWM in the presence of cord blood T cells (Morito et al. 1979; Tosato et al. 1980; Andersson et al. 1980). In addition, EBV-stimulation of cord B cells is abrogated by coculture with cord T cells and PWM, suggesting that cord blood, in contrast to adult blood, contains a subpopulation of suppressor T cells that are susceptible to activation by PWM. This finding suggests that the suppressor population is functionally immature and requires activation by PWM for suppression to occur. Suppressor T cell activity in cord blood is without preference for immunoglobulin isotype and is completely abolished by irradiation or hydrocortisone, indicating that continued suppression is dependent on cell division. Miyawaki et al. (1981) has demonstrated that suppressive activity is related to elaboration of a soluble suppressor factor, termed cord T-cell-derived suppressor factor (CTSF). Functional studies suggest that CTSF-mediated suppression occurs via collaboration with monocytes.

7 Conclusions

A simplified interpretation of the data available on neonatal immunity is that normal numbers of B and T lymphocytes are present at birth, yet the newborn's humoral immune responses are functionally immature. This paradoxical relationship between immunological repertoire and functional capability has been substantiated through clinical observations of congenitally infected infants whose predominant antibody response is limited to IgM, by the neonate's deficient antibody production to capsular polysaccharide antigens and by the delayed acquisition of IgG and IgA isotypes. Information derived from in vitro studies has led to speculation that regulatory imbalances between T-cell-mediated help and suppression, as well as an intrinsic B-cell immaturity, may be responsible for the relative immunodeficiency of the newborn.

It is tempting to speculate that the gap in the humoral responses of the newborn is not the result of an intrinsic B-cell defect but rather a lack of immunological experience. Although the B-cell repertoire is complete by the second trimester of gestation, clonal expansion as a result of exposure to exogenous antigens is unusual during normal fetal development. Consequently, the population of memory B cells in the newborn capable of responding to antigen challenge is low. The bone marrow produces large numbers of B cells each day; however, there is evidence that few of these migrate to peripheral tissues to become part of the established repertoire. Studies by De Freitas and Coutinho

(1981) suggest that newly formed B cells decay at an average rate of 50% per day. Thus, the acquisition of memory B cells is a slow process and may explain the infant's delayed humoral immune responses. In this regard, previous studies suggest that participation of memory B cells is especially necessary for in vitro responses to PWM. The PWM-responsive B-cell subpopulation comprises relatively large, low density cells whose features suggest in vivo preactivation (Dagg et al. 1981; Findlay et al. 1979). These B cells reflect prior antigen exposure of the donor, a finding that may be related to the delayed differentiation responses of neonatal B cells to PWM. Thus, the lag in the development of humoral immunity may reflect a lack of previous experience with exogenous antigens rather than an intrinsic B-cell defect.

Remaining controversies concerning the mechanisms of functional immaturity of the newborn assure continued interest in this problem. Newer approaches, such as cell cycle analysis, delineation of sequential expression of surface activation antigens by monoclonal antibodies, and responsiveness to lymphokines, offer great promise in the further elucidation of neonatal B-cell activation and differentiation.

References

Abney ER, Cooper MD, Kearney JF, Lawton AR, Parkhouse RME (1978) Sequential expression of immunoglobulin on developing mouse B-lymphocytes. A systematic survey which suggests a model of T and B cells. J Immunol 119:1213-1217

Alford CA Jr, Wu LYF, Blanco A (1974) Developmental humoral immunity and congenital infections in man. In: Neter J, Milgram F (eds) The Immune system and infectious diseases. Karger, Basel, pp 42-58

Alt FW, Yancopoulos GD, Blackwell TK, Wood C, Thomas E, Boss M, Coffman R, Rosenberg N, Tonegawa S, Baltimore D (1984) Ordered rearrangement of immunoglobulin heavy chain variable region segments. EMBO J 3:1209-1219

Andersson U, Bird G, Britton S (1980) Cellular mechanisms of restricted immunoglobulin formation in the human neonate. Eur J Immunol 10:888-894

Andersson U, Bird AG, Britton S (1981) Humoral and cellular immunity in humans studied at the cell level from birth to two years of age. Immunol Rev 57:5-37

Bankhurst AD, Warner NL (1972) Surface immunoglobulins on the thoracic duct lymphocytes of the congenitally athymic (nude) mouse. Aust J Exp Biol Med Sci 50:661-664

Butler JL, Muraguchi A, Lane HC, Fauci AS (1983) Development of a human T-T cell hybridoma secreting B cell growth factor. J Exp Med 157:60-68

Cooper MD, Peterson RDA, South MA, Good RA (1966) The functions of the thymus system and the bursa system in the chicken. J Exp Med 123:75-102

Dagg MK, Levitt D (1981) Human B lymphocyte subpopulations. I. Differentiation of density-separated B lymphocytes. Clin Immunol Immunopathol 21:39-49

Davis MM, Kim SK, Hood LE (1980) DNA sequences mediating class switching in α immunoglobulins. Science 209:1360-1365

De Freitas AA, Coutinho A (1981) Very rapid decay of mature B lymphocytes in the spleen. J Exp Med 994-999

Findlay GJ, Both RJ, Marbrook J (1979) Antibody responses of human lymphocytes in vitro. Specificity and physical properties of plaque forming cell precursors. Aust J Exp Biol Med Sci 57:587-596

Forni L, Coutinho A (1982) The production of membrane or secretory forms of immunoglobulins is regulated by C-gene-specific signals. Nature 299:173-175

Fowler R, Shubet WK, West CD (1960) Acquired partial tolerance to homologous skin grafts in the human infant at birth. Ann NY Acad Sci 87:403-428

Gathings WE, Lawton AR, Cooper MD (1977) Immunofluorescent studies of the development of pre-B cells, B lymphocytes and immunoglobulin isotype diversity in humans. Eur J Immunol 7:804–810

Gillies DS, Morrison KSL, Oi VT, Tonegawa S (1983) A tissue-specific transcription enhancer element is located in the major intron of a rearranged immunoglobulin heavy chain gene. Cell 33:717–728

Hayward AR, Lawton AR (1977) Induction of plasma cell differentiation of human fetal lymphocytes: evidence for functional immaturity of T and B cells. J Immunol 119:1213–1217

Hayward AR, Kurnick J (1981) Newborn T cell suppression: early appearance, maintenance in culture, and lack of growth factor suppression. J Immunol 126:50–53

Hieter PA, Korsemeyer SJ, Waldmann TA, Leder P (1981) Human immunoglobulin $\varkappa$ light-chain genes are deleted or rearranged in λ-producing B cells. Nature 290:368–372

Honjo T (1983) Immunoglobulin genes. Annu Rev Immunol 1:499–528

Hood L, Kronenberg M, Hunkapiller T (1985) T cell antigen receptors and the immunoglobulin supergene family. Cell 40:225–229

Kamps WA, Cooper MD (1982) Microenvironmental studies of pre-B and B cell development in human and mouse fetuses. J Immunol 129:526–531

Kishimoto T (1985) Factors affecting B-cell growth and differentiation. Annu Rev Immunol 3:135–157

Kiyono H, Cooper MD, Kearney JF, Mosteller LM, Michalek SM, Koopman WJ, McGhee JR (1984) Isotype-specificity of helper T cell clones: Peyer's patch Th cells preferentially collaborate with mature IgA B cells in IgA responses. J Exp Med 159:798–811

Konowalchuk J, Speirs JI, Perelmutter L (1982) Immunoglobulin properties of Epstein-Barr virus transformed human umbilical cord and adult peripheral blood lymphocytes. Cell Immunol 190–196

LeDouarin NM, Jotereau FV (1975) Tracing of cells of the avian thymus through embryonic life in interspecific chimeras. J Exp Med 142:17–40

Mayer L, Posnett DN, Kunkel HG (1985) Human malignant T cells capable of inducing an immunoglobulin class switch. J Exp Med 161:134–144

Mayumi M, Kuritani T, Kubagawa H, Cooper MD (1983) IgG subclass expression by human B lymphocytes and plasma cells: B lymphocytes precommitted to IgG subclass can be preferentially induced by polyclonal mitogens with T cell help. J Immunol 130:671–677

Mills EL, Regelmann WE, Quie PG (1981) Immunology of the newborn. In: Feigin RD, Cherry JD (eds) Textbook of pediatric infectious diseases. Saunders, Philadelphia, p 738

Miyawaki T, Seki H, Kubao M, Taniguchi N (1979) Suppressor activity of T lymphocytes from infants assessed by co-culture with fractionated adult lymphocytes in the pokeweed mitogen system. J Immunol 123:1092–1096

Miyawaki T, Moriya N, Nagaoki T, Kubo M, Yokoi T, Taniguchi N (1981) Mode of action of humoral suppressor factor derived from pokeweed mitogen-stimulated cord T cells on adult B cell differentiation. J Immunol 126:282–285

Moretta L, Mingari MC, Sekaly PR, Moretta A, Chapius B, Cerotini J (1981) Surface markers of cloned human T cells with various cytolytic activities. J Exp Med 154:569–574

Morito T, Bankhurst AD, Williams RC, Jr (1979) Studies of human cord blood and adult lymphocyte interactions with in vitro immunoglobulin production. J Clin Invest 64:990–995

Norris EH (1938) The morphogenesis and histogenesis of the thymus gland in man. Contrib Embryol 27:193–207

Oldstone MBA, Tishon A (1977) Active thymus derived suppressor lymphocytes in human and cord blood. Nature 269:333–335

Osmond DG, Nossal GJV (1974) Differentiation of lymphocytes in mouse bone marrow. II. Kinetics of maturation and renewal of anti-globulinbinding cells studies by double labeling. Cell Immunol 13:132–145

Parkman R, Mosier D, Umansky, Cochran W, Carpenter CB, Rosen FS (1974) Graft-versus-host disease after intrauterine and exchange transfusions for hemolytic disease of the newborn. N Engl J Med 290:359–363

Reinherz EL, Kung PC, Goldstein G, Levey RH, Schlossman SF (1980) Discrete stages of human intrathymic differentiation: Analysis of normal thymocytes and leukemic lymphoblasts of T-cell lineage. Proc Natl Acad Sci USA 77:1588–1592

Romangnani S, Guidizi MG, Biagiotti R, Almerigogna F, Maggi E, Del Prete G, Ricci M (1981) Interaction of SPA with surface IgG. J Immunol 127:1307–1313

Rossant J, Vijh KM, Grossi CE, Cooper MD (1986) Clonal origin of haemopoietic colonies in the postnatal mouse liver. Nature (in press)
Ruuskanen O, Pittard WB, Miller K, Pierce G, Sorensen RU, Polman SH (1980) *Staphylococcus aureus* Cowan I-induced immunoglobulin production in human cord blood lymphocytes. J Immunol 125:411–413
Suzuki T, Sanders SK, Butler JL, Cooper MD (1986) Identification of a novel differentiation antigen on activated human B cells (to be published)
Thomas Y, Rogozinski L, Irigoyen OH, Friedman SM, Kung PC, Goldstein G, Chess L (1981) Functional analysis of human T cell subsets defined by monoclonal antibodies: IV. Induction of suppressor cells within the $OKT4^+$ population. J Exp Med 154:459–467
Tonegawa S (1983) Somatic generation of antibody diversity. Nature 302:575–581
Tosato G, Magrath IT, Koski IR, Dooley NJ, Blaese RM (1980) B cell differentiation and immunoregulatory T cell function in human cord blood lymphocytes. J Clin Invest 66:383–388
Wabl M, Burrows PD (1984) Expression of immunoglobulin heavy chain at a high level in the absence of a proposed immunoglobulin enhancer element in cis. Proc Natl Acad Sci USA 81:2452–2455

Regulation of Antibody Synthesis in the Neonate

ULF ANDERSSON

Dept. of Immunobiology, Karolinska Institute, Wallenberglaboratory Lilla Frescati, 104 05 Stockholm, Sweden

1 Introduction

The humoral immune response is reduced during the neonatal period. In my opinion, this is mainly due to immature cell functions and a paucity of stimulation by exogenous antigens. The widely accepted concept that active suppression contributes to this relative immunoincompetence will be challenged in this article.

The serum level of IgG in the full-term neonate is at least equal to that of an adult. This maternally derived IgG provides the infant with a good protection against many infections, but it complicates our efforts to define the endogenous capacity of the newborn to defend himself against various antigens. Whether the transferred IgG exert any positive or negative influence on the maturation of the fetal immune system remains an unanswered question. The normal immunologic development of the offspring of agammaglobulinemic mothers indicates (Holland and Holland 1966; Leuersen and Christensen 1973; Kobayashi et al. 1980) that maternal IgG is not an absolute requirement for this process.

Antibody formation depends on an intimate cooperation between T lymphocytes and antigen-presenting B lymphocytes and macrophages, with the exception of certain thymus-independent antigens, which appear to trigger B cells directly. To aid in the understanding of antibody responsiveness in the neonatal period, I shall first summarize what can be learned from in vivo observations about the development of the above cells during fetal life and then go on to discuss their functional capacity as revealed by in vitro studies.

2 Ontogeny of B Lymphocytes

The study of aborted fetuses has provided us with information about most lymphoid organs up to midgestation. Our knowledge about the further development is, however, mainly derived from umbilical cord blood examinations, which leaves us with an unsatisfactory gap in our knowledge.

Since the genetic basis for B lymphocyte differentiation is already outlined in this volume (Cooper), I refer to the latter chapter. The ontogenetic development of the B-cell system has been reviewed excellently by the same author and his coworkers (Gathings et al. 1981), and this work will be cited in short here.

Immunology of the Neonate
Burgio, Hanson, Ugazio (Eds.)

Pre-B cells containing μ chains and, somewhat later, also light chains appear in the fetal liver during the 8th week at first and soon after in the bone marrow. Since these cells do not display any surface Ig (sIg) molecules, it can be concluded that antibody diversity is generated without the influence of environmental antigens. sIgM$^+$ B cells are generated within a week after the pre-B cells are produced. From 12 weeks onwards the bone marrow gradually takes over from the fetal liver as the main source of cells of the B-cell lineage. B cells start to express both sIgM and sIgD during the 4th month and subpopulations coexpress sIgG or sIgA in addition. Although adult proportions of IgM-, IgG-, and IgA-bearing cells are present well before midgestation, the Ig secreted in response to infections is mainly restricted to IgM throughout fetal life. This discrepancy could be explained by postulating that expression of sIg is regulated by RNA splicing processes, while a switch of the secreted IgM to another isotype requires rearrangements at the DNA level, and that these translocation events are delayed during fetal life. IgM-containing plasma cells have been found in response to infections beyond the 15th week, and small numbers of IgG- and IgA-positive plasma cells have been reported after the 20th and the 30th weeks, respectively.

B cells constitute 15%–20% of the mononuclear cells in umbilical cord blood at full term. Considering that there are roughly twice as many lymphocytes in cord blood as in peripheral blood from adults, this means that there are many more B cells in the circulation during the neonatal period than later in life (Thomas and Linch 1983). The majority of the neonatal blood B cells express both sIgM and sIgD (Gandini et al. 1981). Surface IgG$^+$ or sIgA$^+$ B cells coexpress IgM and IgD, in contrast to those found in adults, whose IgG- or IgA-bearing B cells generally only display one isotype at the cell surface. Other markers of immaturity characteristic of cord blood B cells are a low ecto-5'-nucleotidase activity (Rowe et al. 1980) and an increased capacity to bind *Helix pomatia* agglutinin (Hellström et al. 1978) and mouse erythrocytes (Forbes et al. 1981).

3 In Vivo Ig Formation During Neonatal Life

The newborn seems capable of mounting an Ig response in vivo to a variety of antigens (Fink et al. 1962; Uhr et al. 1962; Smith et al. 1964; Pabst and Kreth 1980), which is strongly dominated by IgM but also comprises small amounts of IgG1 and IgG3 (Oxelius and Svenningsen 1984). However, by analogy to what is found in the murine model, there seems to be an inability to respond to certain antigens, mainly of the polysaccharide group, at this stage of life (South 1972; Gold et al. 1975; Cowan et al. 1978). From studies in inbred animals, it is obvious that the B cell antigen repertoire follows a strictly predetermined sequential program and that in this antigen hierarchy the polysaccharide group is one of the last to elicit an immune response (D'Eustachio et al. 1977; Fernandez and Möller 1977).

When Smith et al. (1964) immunized newborn humans with *Salmonella* strains to which they lacked maternal antibodies, informative differences from the adult controls were noted. A strong response was evoked to the protein H antigen in the newborns, but the antibody production remained of the IgM class during the 1st month after vaccination, while the immunized adults switched from IgM to IgG production within 5–15 days. The polysaccharide O antigens did not induce any antibody formation at all in the newborns. An excellent immune response of neonates to hepatitis B vaccine has been reported recently by Lee et al. (1983). In contrast, other studies have found a diminished antibody response of neonates to diphtheria toxoid (Vahlquist 1949; Dancis et al. 1953) to inactivated poliomyelitis vaccine (Perkins et al. 1958) and to pertussis vaccine (Provenzano et al. 1965). Taken together, these in vivo observations suggest a reduced antibody responsiveness in the newborn to certain but not other antigens, and a delayed ability to switch from IgM to IgG production.

4 Ontogeny of T Lymphocytes

Stem cells enter the thymus at the end of the 2nd month. It is now possible to characterize the phenotype of the differentiating thymocytes during gestation using monoclonal antibodies. These tools have shed new light on T cell function during fetal and neonatal life, since many of the antigens detected by the antibodies are correlated to functional properties of the T cells. A differentiation scheme for individual cells in the thymus, based on reactivity with monoclonal OKT antibodies, has been proposed by Reinherz et al. (1980): The earliest lymphoid cells in the thymic cortex lack mature T-cell antigens with the exception of T11, the sheep erythrocyte receptor, but express T9 and T10 antigens (stage 1). With further differentiation thymocytes lose T9, retain T10, and acquire the thymocyte-specific antigen T6, and concurrently coexpress the T4 and T8 antigens (stage 2). During maturation, the thymocytes then lose the T6 antigen, express T3, and segregate to $T4^+/T8^-$ and $T4^-/T8^+$ subsets, respectively (stage 3). Immunological competence is not fully developed until the thymic lymphocytes leave the thymus. Resting peripheral T cells in adults no longer express the T10 antigen.

When Asma et al. (1983) used these reagents to study the ontogenetic development, they found maturation stages in fetal thymus (11–20 weeks) comparable to those in postnatal life, with the exception that a great proportion of fetal thymocytes expressed T3 and T6 antigens simultaneously. Forty percent of fetal spleen lymphocytes (12–20 weeks) were T3 positive, and most were also T10 positive, indicating phenotypical immaturity.

Many groups have reported somewhat reduced proportions of T3- and T11-bearing lymphocytes in cord blood at full term compared to peripheral blood from adults. In addition, it has been claimed that the density of the T11 antigen is decreased on cord blood T cells (Rodriguez et al. 1981), which is of interest since both T3 (Meuer et al. 1983) and T11 (Meuer et al. 1984) are receptors for activation signals. There are no remarkable differences in the proportions of T4- and T8-bearing cells in neonates versus adults, and there is no evidence for great number of cells coexpressing these antigens. However, a striking discovery was that a majority of the cord blood T cells still express the T10 antigen (Foa et al.

1984; Gerli et al. 1984; G Griffiths-Chu et al. 1984), suggesting that a large pool of the neonatal T cells represent an intermediate stage between thymocytes and mature peripheral T cells. The presence of T6-bearing cells and $T3^-/T8^+$ cells (Maccario et al. 1983) in cord blood is also in contrast to adult controls. Very few blood T cells from newborns, like those from adults, display HLA-D antigens or IL2 receptors, which is a strong argument against the possibility that T10 expression is a consequence of in vivo activation. These observations suggest that a large fraction of the circulating T lymphocytes in the neonate leave the thymus as immature cells and have to complete their maturation in the periphery, in the absence of thymic humoral influence.

5 Polyclonal B-Cell Activation In Vitro

Cell proliferation and polyclonal production of all Ig isotypes are usually seen when lymphocytes from adults are cultured in vitro and exposed to either T-cell-dependent or independent mitogens. Such activation of cord blood lymphocytes has led to distinctly divergent results from those found in lymphocytes from adults. Although cell proliferation is equal to, or higher than, that in cell cultures of adults, Ig responses are mainly restricted to IgM production, and T-cell-dependent Ig formation is strikingly reduced.

Live Epstein-Barr virus (EBV) is a direct B-cell activator which induces a strong early IgM production in fetal liver B cells from midgestation (Pereira et al. 1982). Cord blood B cells make IgM as do B cells from peripheral blood from adults after EBV stimulation, but not IgG or IgA (Andersson et al. 1981). Since mitogens activate discrete subsets of B cells which differ in degree of differentiation, it is of interest to compare results obtained with polyclonal activators other than EBV.

Killed *Staphylococcus aureus* Cowan type I (SACl) is also a T-independent mitogen (Forsgren et al. 1976), which may induce production of IgM, but not IgG, equally well in blood B cells from neonates and adults (Romagnani et al. 1980; Ruuskanen et al. 1980; Hayward 1981). Occasional cord blood cultures producing some IgA have been reported by Ruuskanen et al. (1980).

Nocardia opaca water-soluble mitogen (NWSM), which may stimulate B cells directly but requires T-cell help for optimal activation, generates some IgM production in cord blood cells (Miyawaki et al. 1981). An adult type of response is, however, noted in blood lymphocytes from infants 1 month of age but is still mainly restricted to IgM secretion.

Most of the in vitro studies with neonatal cells have been performed with pokeweed mitogen (PWM) stimulation, which is a T-cell-dependent process (Keightly et al. 1976). OKT4-positive T cells collaborate with monocytes and B cells, leading to plasma cell differentiation. This response can be inhibited by $OKT8^+$ T cells, also activated by PWM. Furthermore, the $OKT4^+$ T cells can be subgrouped using a recently described monoclonal antibody called 4B4 (Morimoto et al. 1985). The $T4^+4B4^+$ subset provides the helper signal for PWM-induced Ig synthesis, while $T4^+4B4^-$ cells function as inducers of $T8^+$ suppressor cells. The cellular cooperation in the PWM system does not seem to be restricted by HLA differences (Keightly et al. 1976).

All studies performed with PWM-stimulated mononuclear cells from cord blood at full term or from fetal spleen agree that, despite a strong proliferative response, Ig production is minimal (Hayward and Lawton 1977; Morito et al. 1979; Tosato et al. 1980; Andersson et al. 1981; Miyawaki et al. 1981). In our hands, roughly 10% of the PWM-activated cultures from newborn blood donors will lead to Ig secretion, which is composed exclusively of IgM and is maximally 25% of that seen in cultures from adults. The addition of T cells from adults greatly enhances the IgM PWM-driven responses of neonatal B cells while the IgG and IgA plasma cell responses remain deficient (Tosato et al. 1980; Andersson et al. 1981; Yachie et al. 1981). T cells from cord blood, on the other hand, do not show any appreciable facilitation of B cell differentiation when cocultured with non-T cells from adults (Hayward and Lydyard 1979; Andersson et al. 1981; Miyawaki et al. 1981; Tan and Shore 1984), even when OKT4-enriched cell preparations are added (Yachie et al. 1981; Andersson et al. 1983). Taken together, these results would argue that a deficient helper T-cell function and an immature B-cell function restrict IgG and IgA plasma cell responses. Another fact that suggests reduced helper T-cell capacity is that less than 5% of the cord blood T cells express HLA-D antigens as a sign of activation after PWM culture, while more than 40% of T cells (mainly $OKT4^{+}$) from adults do so (Miyawaki et al. 1982). I assume that the $OKT4^{+}4B4^{+}$ subset in cord blood is either absent, or functionally immature. The 10% of cord blood cultures that respond to PWM stimulation with IgM production would then reflect a more advanced stage of T-cell maturation than the remaining majority.

Very interesting results have been reported by Miller et al. (1984), who combined PWM and the thymus-independent mitogen SACl to stimulate cord blood cultures. None of these activators alone led to Ig secretion. However, when combined in a specific sequence they elicited a very strong Ig response of undefined isotype in the cord blood cultures. They also found that conditioned media prepared from activated monocytes from adult donors could replace the effect of the SACl stimulation, and that conditioned media from activated T cells from adults could substitute for the action of PWM. The same results were obtained with the substitution of adult monocytes for cord monocytes or of adult T cells for cord T cells. These data may suggest a functional immaturity of cord blood T cells, as well as of monocytes, in ability to deliver the signals for T-cell-dependent B-cell differentiation.

Other T-cell-dependent routes that have failed to induce Ig synthesis in cord blood B cells include PHA stimulation (Hayward et al. 1981) and autologous mixed lymphocyte reaction (AMLR) (Palacios and Andersson 1982). AMLR-derived supernatants from cultures from adults can, in contrast to those from newborns, support a strong IgM plasma cell response in PWM-activated cord blood cultures.

The above results notwithstanding, there are also examples of successful T-cell-dependent stimulation of neonatal B cells to Ig secretion. Lipopolysaccharide, which is a T-cell-dependent activator with peripheral blood cells (Levitt et al. 1981), induced IgM formation equally well in cultures from newborns as from adults (Andersson et al. 1981). Glucocorticosteroids (GCS) are polyclonal activators of Ig production requiring functional T cells and monocytes in vitro (Grayson et al. 1981), and all isotypes are produced in cultures from adults. Around 30% of the cultured cord blood samples made IgM after GCS stimulation, and this figure improved to 60% when the number of T cells was raised in the cultures (Orson et al. 1984). Additional adult T-cell help to the neonatal cells increased the IgM production and also led to some IgG and IgA formation. This may be a consequence of GCS-induced maturation since these hormones are known to enhance differentiation in many different fetal tissues.

In order to find out how long the neonatal pattern of limited responsiveness persists, I have cultured blood lymphocytes from healthy children of various ages and assessed Ig production after PWM or EBV stimulation (Andersson 1985). The results demonstrate a gradual maturation process, with no abrupt change when the fetomaternal relationship is terminated at delivery.

6 Antigen-Induced Ig Formation In Vitro

Antigen-specific activation of neonatal B cells has been unsuccessful until recently. One factor contributing to the absence of an in vitro antibody response is a paucity in cord blood of precursor B cells for certain antigens as influenza virus (Yarchoan and Nelson 1983) or the autoantigens thyroglobulin and IgG (Fong et al. 1981), but not for TNP-*Brucella abortus* (TNP-Ba) (Golding et al. 1984). TNP-Ba has been classified as a T-independent type-I antigen in the mouse, since it activates neonatal and CBA/N murine B cells, in contrast to T-independent type-2 antigens (Mond et al. 1978). Also human cord blood B cells could be activated in a T-cell-independent manner by this antigen, although the response was lower than that of cultures from adults (Golding et al. 1984).

The T-dependent antigen, inactivated type-A influenza virus, induced antigen-specific antibodies in cultures with blood cells from almost all adults, but failed to do so in cord blood cultures (Yarchoan and Nelson 1983). Newborns had a decreased number of precursor B cells committed to anti-influenza virus Ig production and their T cells did not proliferate in response to the virus. Pascal et al. (1980) used the T-dependent antigen TNP polyacrylamide beads and found Ig formation in 10% of the cord blood samples.

Very exciting results have recently been reported by Van Tol et al. (1984b). When they stimulated cord blood mononuclear cells with six different T-cell-dependent antigens separately, they obtained antigen-specific IgM production equal to that in cultures from adult controls, with the exception of one antigen. Furthermore, the antigen dose optimal for the induction of a response was 100–1000 times lower in cord blood cultures than in the controls. The dose-response difference was explained by an altered handling of the antigens by the monocytes from the neonates. The helper function for IgM formation was obviously far more advanced in these experiments than that revealed after polyclonal T-cell activation, for reasons which are obscure. However, the results of Van Tol and his group resemble, in my opinion, much more those which could be expected from in vivo observations during late gestation and the neonatal period, when IgM is formed in response to many, but not all, infectious agents. It will be of great interest to see whether these results are confirmed in other laboratories.

One objection to all the experiments so far discussed concerns the culture conditions. The fetal or neonatal cells have always been cultured in media supplemented with serum from adult individuals or occasionally in xenogeneic serum, while their performance in autologous serum has never been described to my knowledge.

7 Macrophage Function

Myeloid progenitor cells appear early during ontogeny mainly in the fetal bone marrow (Rosenthal et al. 1983) but we do not have much information about their functional capacity during gestation. Macrophages regulate Ig formation by providing interleukin 1 (IL1) and by presenting the antigen to the helper T cells (Scala and Oppenheim 1983). Cord blood monocytes have been demonstrated to be good producers of IL1 (Zlabinger et al. 1983). Cord blood monocytes express significant amounts of HLA DR and DP, although in somewhat lower density than monocytes from peripheral blood from adults (Edwards et al. 1985; Van Tol et al. 1984a). DQ molecules are, however, displayed by a much lower number of monocytes in cord blood than in the controls, and there seems to be a delayed appearance of DQ products during the ontogenetic development of other fetal tissues as well (Edwards et al. 1985).

According to Gonwa et al. (1983), one could expect a reduced capacity for presentation of certain antigens as a consequence of low HLA-DQ expression, but so far this has not been found in the few experiments dealing with antigen presentation by cord blood monocytes (Hoffman et al. 1981; Zablinger et al. 1983; Van Tol et al. 1984a). Hoffman et al. (1981) found that tetanus-toxoid-specific T cell blasts from maternal blood continued to proliferate in cocultures with semiallogeneic cord blood mononuclear cells pulsed with tetanus toxoid. The successful results using cord blood cultures stimulated with T-cell-dependent antigens have already been described (Van Tol et al. 1984b). The remarkable finding that neonatal monocytes needed so much less antigen for efficient presentation cannot be explained at present. Prostaglandin E_2 (PGE_2) is ascribed a decisive role by the authors, but since it has been reported that cord blood mononuclear cells are extremely resistant (Johnsen et al. 1983), as well as excessively sensitive, to PGE_2 effects (Wakasugi et al. 1985), it seems premature to draw any firm conclusions. We need to know much more about how antigen is handled by macrophages in the neonatal period to understand how antibody formation is regulated.

8 Helper T-Cell Function

Cord blood T cells are fully mature in their capability to proliferate and respond to mitogens and alloantigens, and in autologous mixed lymphocyte reaction, which all indicate normal interleukin 2 (IL2) production (Hayward 1981; Palacios and Andersson 1982). These T cells are also able to support cell-mediated lysis of allogeneic target cells (Granberg et al. 1979). Preterm infants born at 32–36 weeks of gestation show tuberculin conversion following BCG vaccination at birth (Dawodu 1985). These facts demonstrate that many T-cell functions are fully mature at birth.

It is evident from data presented previously that it is much more unclear whether the helper T-cell function for Ig production is fully developed or not at full term, since the in vitro results with mitogen and antigen stimulation are so divergent. Cord blood T cells, which support mitogen-induced B-cell differentiation poorly, will generally do so after irradiation (1000–2500 rad) but not to the functional level of T cells from adults (Hayward and Lydyard 1979; Tosato et al. 1980; Pedersen et al. 1983). These effects have been ascribed to an elimination of suppressor T-cell influence and will be dealt with later.

The mechanisms by which T cells collaborate with B cells in the generation of specific antibody responses involve a direct cell interaction, which is antigen and Ia restricted (Lanzavecchia 1985), and the delivery of various lymphokine signals which influence B-cell activation, proliferation and differentiation. These T-cell factors, such as B-cell growth factor(s) (BCGF), B-cell differentiation factor(s) BCDF, isotype-switch factor(s), and possibly others (reviewed by Möller 1984), are still poorly characterized. Even less is known about the production of these lymphokines in the newborn.

Two well-characterized T-cell products, IL2 and interferon-γ (IFN-γ), have been shown to mediate BCGF and BCDF activity, respectively, when added together in cultures of SACl-stimulated B cell blasts (Nakagawa et al. 1985). The neonatal production of IL2 and IFN-γ has been analyzed. Cord blood T cells make IL2 in abundance after PHA, ConA, or OKT3 stimulation, but remarkably no IFN-γ (Wakasugi and Virelizier 1985; Taylor and Bryson 1985; personal unpublished data). The neonatal T cells are, however, not incapable of IFN-γ production, since when they were provided with another stimulus such as staphylococcal enterotoxin A (SEA), large amounts fully comparable to that from adult controls were made (Wakasugi and Virelizier 1985). Whether SEA activates different subsets of producer cells or uses a different activation pathway than the other mentioned mitogens is unknown. These data may indicate that there is an asynchronous ontogenetic onset for production of the lymphokines involved in T-cell regulation of Ig formation.

We have tested experimentally whether the passive addition of recombinant IFN-γ to PWM-stimulated cultures effects the subsequent Ig production. A considerable increase was noted consistently in cultures from adult donors (Andersson et al. 1985) and also in those rare cord blood samples that produced Ig in response to PWM stimulation, while all the neonatal cultures which did not make Ig remained nonresponders after addition of IFN-γ (unpublished data). Thus, we could not reproduce the synergistic effects seen with PWM and SACl on Ig production (Miller et al. 1984) by combining PWM and IFN-γ, as we had hoped. Nevertheless, I find it an interesting coincidence that these staphylococcal products reveal otherwise disguised properties of cord blood T cells, such as a capacity for IFN-γ production and PWM-induced Ig synthesis.

9 Suppressor T Cell Function

The observations that cord blood T cells inhibit alloantigen- and mitogen-induced cell proliferation (Olding et al. 1974; Lawler et al. 1975) and Ig production of cells from adults in cocultures (Durandy et al. 1979; Tosato et al. 1980; Andersson et al. 1981; Hayward 1981; Miyawaki et al. 1981), and that they poorly sustain Ig synthesis in autologous cultures activated with T-dependent mitogens, suggest a role of suppressor mechanisms. It has been proposed that the T-cell system in the neonate is functionally balanced towards suppression. However, this cannot be a general rule for the endogenous regulation, since many T-and B-cell functions appear to be fully normal at birth. One remarkable feature of the neonatal T-cell-mediated suppression is that it does not affect alloantigen- or mitogen-stimulated cell division (Olding and Oldstone 1976; Olding et al. 1977), or Ig synthesis (Andersson et al. 1981, 1983) of allogeneic neonatal cells, in contrast to allogeneic cells from adults. We found that cord blood T cells inhibited EBV-induced Ig secretion in cocultures with mononuclear cells from adults by releasing IFN-γ, and that IFN-γ in synergy with IFN-α exerted a protective influence on B cells from adults, while B cells from neonates were totally insensitive to these antiviral effects (Andersson et al. 1983; 1985). Although EBV stimulation cannot be regarded as a representative model to study the normal regulation of Ig production, I think that the different effects seen at the target cell level between adults versus neonates is of fundamental interest.

We also found that IFN-γ was not involved in the neonatal T-cell-mediated inhibition of Ig production in PWM-stimulated cultures from adults, a fact that argues against the presence of a common suppressor T-cell function during fetal life (Andersson et al. 1983).

The existence of neonatal suppressor T cells acting on helper T cells needed to support plasma-cell differentiation is still a possibility. This would be difficult to detect in the allogeneic culture systems described so far. This suppression is hardly a generalized phenomena in any case, since many T-cell-dependent antigens and some mitogens seem to elicit a good IgM response in cord blood cultures. Furthermore, Orson et al. (1984) were able to enhance Ig responses in GCS-stimulated neonatal cultures by increasing the proportion of T cells to non-T cells, which does not support the idea of a dominant suppressor function in the T-cell compartment. To my mind, the synergistic effect of SAC1 and PWM on Ig secretion in cord blood cultures (Miller et al. 1984) is not explained primarily by a release from suppression.

The main argument for functional suppressor T cells affecting autologous Ig production comes from experiments with PWM. Several groups have identified some helper T-cell function for Ig synthesis after irradiating (1000–2500 rad) the neonatal T cells (Hayward and Lydyard 1979; Tosato et al. 1980; Pedersen et al. 1983). These results have been interpreted as evidence for a concealed helper capacity appearing when the influence of the radiosensitive suppressor T cells is abolished. However, in my opinion, this unphysiological treatment of the cells could affect differentiation of the helper function in a direct manner. Whatever the explanation may be, I think one can conclude that neonatal T cells are not intrinsically suppressive. They may become so as a consequence of PWM stimulation, but not in response to many other mitogens and antigens. I think there are good reasons for hesitation in ascribing functional deficiencies in the neonatal immune system to the action of suppressor T cells.

10 Concluding Remarks

There seems to be an inherent immaturity of B-cell function in the human newborn which restricts Ig responses mainly to the IgM class. The predominance of IgM is evident regardless of whether T-dependent or independent stimuli are used, which makes it less likely that T cells are responsible for the lack of IgG and IgA. Furthermore, all neonatal IgG or IgA surface-positive B cells coexpress IgM and IgD, which represent immature phenotypes rarely displayed by B cells from adults. These neonatal B cells can mount antibody responses to T-dependent and most T-independent antigens, with the exception of a group of polysaccharide antigens. This inability may have deleterious consequences if the neonate lacks protective maternal antibodies when exposed to pathogens. The helper T-cell function for Ig production is presumably well advanced in the neonate as judged from in vivo and in vitro responses to antigens, and it has probably been underestimated in previous experiments using mitogen activation. Suppressor mechanisms do not appear to be a major explanation for reduced immunological responsiveness in the neonate.

References

Andersson JP, Andersson U, Ernberg I, Britton S, De Ley M (1985) Effects of pure interferons on Epstein-Barr virus infection in vitro. J Virol 54:615–618

Andersson U (1985) Development of B lymphocyte function in childhood. Acta Paediatr Scand 74:568–573

Andersson U, Bird AG, Britton S, Palacios R (1981) Humoral and cellular immunity in humans studied at the cell level from birth to two years of age. Immunol Rev 57:5–38

Andersson U, Britton S, De Ley M, Bird G (1983) Evidence for the ontogenetic precedence of suppressor T cell functions in the human neonate. Eur J Immunol 13:6–13

Asma GEM, van den Berg RL, Vossen JM (1983) Use of monoclonal antibodies in a study of the development of T lymphocytes in the human fetus. Clin Exp Immunol 53:429–436

Cowan JM, Ammann AJ, Wara DW, Howie VM, Schultz L, Doyle N, Kaplan M (1978) Pneumococcal polysaccharide immunization in infants and children. Pediatrics 62:721–727

Dancis J, Osborn JJ, Kung HW (1953) Studies of immunology of the newborn infant: IV. Antibody formation in the premature infant. Pediatrics 12:151–156

Dawodu AH (1985) Tuberculin conversion following BCG vaccination in preterm infants. Acta Paediatr Scand 74:564–567

D'Eustachio P, Rutishouser US, Edelman GM (1977) Clonal selection and the ontogeny of the immune response. In: Bourne GH, Danielli JF, Jean KW (eds) International review of cytology, suppl 5. Academic, New York, pp 1-60
Durandy A, Fischer A, Griscelli C (1979) Active suppression of B lymphocyte maturation by two different newborn T lymphocyte subsets. J Immunol 123:2644-2650
Edwards JA, Jones DB, Evans PR, Smith JL (1985) Differential expression of HLA class II antigens on human fetal and adult lymphocytes and macrophages. Immunology 55:489-500
Fernandez C, Möller G (1977) Immunological unresponsiveness to thymus independent antigens: Two functionally different genetic mechanisms of B cell unresponsiveness to dextran. J Exp Med 146:1663-1677
Fink CW, Miller WE Jr, Dorward B, LoSpatullo J (1962) Formation of macroglobulin antibodies. II. Studies on neonatal infants and older children. J Clin Invest 41:1422-1428
Foa R, Giubellino MC, Fierro MT, Lusso P, Ferrando ML (1984) Immature T lymphocytes in human cord blood identified by monoclonal antibodies: A model for the study of the differentiation pathway of T cells in humans. Cell Immunol 89:194-201
Fong S, Tsoukas CD, Frincke LA, Lawrance SK, Holbrook TL, Vaughan JH, Carson DA (1981) Age-associated changes in Epstein-Barr virus-induced human lymphocyte autoantibody responses. J. Immunol 126:910-914
Forbes IJ, Zalewski PD, Valente L, Gee D (1982) Two maturation-associated mouse erythrocyte receptors of human B cells. Clin exp Immunol 47:396-404
Forsgren A, Svedjelund A, Wigzell H (1976) Lymphocyte stimulation by protein A of *Staphylococcus aureus.* Eur J Immunol 6:207-213
Gandini M, Kubagawa H, Gathings WE, Lawton AR (1981) Expression of three immunoglobulin isotypes by individual B cells during development: Implication for heavy chain switching. Am J Reprod Immunol 1:161-163
Gathings WE, Kubagawa H, Cooper MD (1981) A distinctive pattern of B cell immaturity in perinatal humans. Immunol Rev 57:107-126
Gerli R, Rambotti P, Cernetti C, Verlardi A, Spinozzi F, Tabilio A, Martelli MF, Grignani F, Davis S (1984) A mature thymocyte-like phenotypic pattern on human cord circulating T lymphoid cells. J Clin Immunol 4:461-468
Gold R, Lepow ML, Goldschneider I, Draper TL, Gotschlich EC (1975) Clinical evaluation of group A and group C meningococcal polysaccharide vaccines in infants. J Clin Invest 56:1536-1547
Golding B, Muchmore AV, Blaese RM (1984) Newborn and Wiskott-Aldrich patient B cells can be activated by TNP-*Brucella abortus:* Evidence that TNP-*Brucella abortus* behaves as a T-independent type 1 antigen in humans. J Immunol 133:2966-2971
Gonwa TA, Picker LI, Raff HV, Goyert SM, Silver J, Stobo JD (1983) Antigen-presenting capabilities of human monocytes correlates with their expression of HLA-DS, an Ia determinant distinct from HLA-DR. J Immunol 130:706-711
Granberg C, Hirvonen T, Toivanen P (1979) Cell-mediated lympholysis by human maternal and neonatal lymphocytes: mother's reactivity against neonatal cells and vice versa. J Immunol 123:2563-2567
Grayson J, Dooley NJ, Koski IR, Blaese RM (1981) Immunoglobulin production induced in vitro by glucocorticoid hormones. J Clin Invest 68:1539-1547
Griffiths-Chu S, Patterson JAK, Berger CL, Edelson RL, Chu AC (1984) Characterization of immature T cell subpopulations in neonatal blood. Blood 64:296-300
Hayward AR (1981) Development of lymphocyte responses and interactions in the human fetus and newborn. Immunol Rev 57:39-60
Hayward AR, Lawton AR (1977) Induction of plasma cell differentiation of human fetal lymphocytes: evidence for functional immaturity of T and B cells. J Immunol 119:1213-1217
Hayward AR, Lydyard PM (1979) B cell function in the newborn. Pediatrics 64:758-764
Hellström U, Perlmann P, Robertsson ES (1978) Receptors for *Helix pomatia:* A haemagglutinin on a subpopulation of human B cells. Scand J Immunol 7:191-197
Hoffman AA, Hayward AR, Kurnick JT, Defreitos EC, McGregor J, Harbeck RJ (1981) Presentation of antigen by human newborn monocytes to maternal tetanus toxoid-specific T-cell blasts. J Clin Immunol 1:217-221
Holland NH, Holland P (1966) Immunological maturation in an infant of an agammaglobulinemic mother. Lancet 2:1152-1155

Johnsen SA, Olofsson A, Green K, Olding LB (1983) Strong suppression by mononuclear leukocytes from cord blood of human newborns on maternal leukocytes associated with differences in sensitivity to prostaglandin E_2. Am J Reprod Immunol 4:45–49

Keightly RG, Cooper MD, Lawton AR (1976) The T cell dependence of B cell differentiation induced by pokeweed mitogen. J Immunol 117:1538–1544

Kobayashi RH, Hyman CJ, Stiehm ER (1980) Immunologic maturation in an infant born to a mother with agammaglobulinemia. Am J Dis Child 134:942–944

Lanzavecchia A (1985) Antigen-specific interaction between T and B cells. Nature 314:537–539

Lauersen HB, Christensen MF (1973) Immunoglobulins in normal infant born of severe hypoglobulinemic mother. Arch Dis Child 48:646–648

Lawler SD, Ukaljiofa EO, Reeves BR (1975) Interaction of maternal and neonatal cells in mixed lymphocyte cultures. Lancet 2:1185–1187

Lee CYG, Hwang LY, Beasly RP, Chen SH, Lee TY (1983) Immunogenicity of hepatitis B virus vaccine in healthy Chinese neonates. J Infect Dis 148:526–529

Levitt D, Stull DD, Lawton AR (1981) T cell-dependent and independent plasma cell-differentiation induced by *Escherichia coli* lipopolysaccharide in human peripheral blood lymphocytes. Clin Immunol Immunopathol 18:309–321

Maccario R, Nespoli L, Mingrat G, Vitiello A, Ugazio AG, Burgio GR (1983) Lymphocyte subpopulations in the neonate: Identification of an immature subset of OKT8 positive OKT3 negative cells. J Immunol 130:1129–1131

Meuer SC, Acuto O, Hussey RE, Hodgdon JC, Fitzgerald KA, Schlossman SF, Reinherz EC (1983) Evidence for the T3-associated 90 kd heterodimer as the T cell antigen receptor. Nature 303:808–810

Meuer SC, Hussey RE, Fabbi M, Fox D, Acuto O, Fitzgerald KA, Hodgdon JC, Protentis JP, Schlossman SF, Reinherz EL (1984) An alternative pathway of T cell activation: A functional role of the 50 kd T11 sheep erythrocyte receptor protein. Cell 36:897–906

Miller KM, Pittard WB, Sorensen RU (1984) Cord blood B cell differentiation. Synergistic effect of pokeweed mitogen and *Staphylococcus aureus* on in vitro differentiation of B cells from human neonates. Clin Exp Immunol 56:415–424

Miyawaki T, Moriya N, Nagaoki T, Taniguchi N (1981) Maturation of B cell differentiation ability and T cell regulatory function in infancy and childhood. Immunol Rev 57:61–87

Miyawaki T, Yachie A, Nagaoki T, Mukai M, Yokoi T, Uwadana N, Taniguchi N (1982) Expression ability of Ia antigens on T cell subsets defined by monoclonal antibodies on pokeweed mitogen stimulation in early human life. J Immunol 128:11–15

Möller G (1984) B cell growth and differentiation factors. Munksgaard, Copenhagen (Immunology review, vol 78)

Mond JJ, Scher I, Blaese RM, Mosier DE, Paul WE (1978) T-independent responses in B cell-defective CBA/N mice to *Brucella abortus* and to trinitrophenyl (TNP) conjugates of *Brucella abortus*. Eur J Immunol 8:459–463

Morimoto C, Letvin NL, Boyd AW, Hagan M, Brown HM, Kornachi MM, Schlossman SF (1985) The isolation and characterization of the human helper inducer T cell subset. J Immunol 134:3762–3769

Morito T, Bankhurst AD, Williams RC (1979) Studies of human cord blood and adult lymphocyte interactions with in vitro immunoglobulin production. J Clin Invest 64:990–995

Nakagawa T, Hirano T, Nakagaawa N, Yoshizaki K, Kishimoto T (1985) Effect of recombinant IL-2 and IFN-γ on proliferation and differentiation of human B cells. J Immunol 134:959–966

Olding LB, Oldstone MBA (1976) Thymus-derived peripheral lymphocytes from human newborns inhibit division of their mothers' lymphocytes. J Immunol 116:682–686

Olding LB, Benirschke K, Oldstone MBA (1974) Inhibition of mitosis of lymphocytes from human adults by lymphocytes from human newborns. Clin Immunol Immunopathol 3:79–89

Olding LB, Murgita RA, Wigzell H (1977) Mitogen-stimulated lymphoid cells from human newborns suppress the proliferation of maternal lymphocytes across a cell-impermeable membrane. J Immunol 119:1109–1114

Orson FM, DeSean V, Pike S, Blaese RM (1984) Glucocorticosteroids stimulate polyclonal immunoglobulin production by cord blood mononuclear cells. J Immunol 133:208–212

Oxelius VA, Svenningsen NW (1984) IgG subclass concentrations in preterm neonates. Acta Paediatr Scand 73:626–630

Pabst HF, Kreth HW (1980) Ontogeny of the immune response as a basis of childhood disease. J Pediatr 97:519–534

Palacios R, Andersson U (1982) Autologous mixed lymphocyte reaction in human cord blood lymphocytes. Cell Immunol 66:88–98

Pascal CP, Galanoud P, Dormont J, Wallon C (1980) Primary in vitro antibody response in human cord blood lymphocytes. J Reprod Immunol 1:275–283

Pedersen SA, Petersen J, Andersen V (1983) Suppression of B lymphocytes in mature newborn infants. Acta Paediatr Scand 72:441–447

Pereira S, Webster D, Platts-Mills T (1982) Immature B cells in fetal development and immunodeficiency: studies of IgM, IgG, IgA and IgD production in vitro using Epstein-Barr virus activation. Eur J Immunol 12:540–546

Perkins FT, Yetts R, Gaisford W (1958) Serological response of infants to poliomyelitis vaccine. Br Med J 2:68–71

Provenzano RW, Wellerlow LH, Sullivan CL (1965) Immunization and antibody response in the newborn infant: I. Pertussis inoculation within twenty-four hours of birth. N Engl J Med 273:959–965

Reinherz EL, Kung PC, Goldstein G, Levey RH, Schlossman SF (1980) Discrete stages of human intrathymic differentiation: analysis of normal thymocytes and leukemic lymphoblasts of T cell lineage. Proc Natl Acad SCI USA 77:1588–1592

Rodriguez MA, Bankhurst AD, Ceuppens JL, Williams RC Jr (1981) Characterization of the supressor cell activity in human cord blood lymphocytes. J Clin Invest 68:1577–1585

Romagnani S, Del Prete GF, Maggi E, Falagiani P, Ricci M (1980) T cell-independence of immunoglobulin synthesis by human peripheral blood lymphocytes stimulated with SpA-containing *Staphylococci*. Immunology 41:921–928

Rosenthal P, Rimm IJ, Umiel T, Griffin JD, Osathanondh R, Schlossman SF, Nadler LM (1983) Ontogeny of human hematopoietic cells: Analysis utilizing monoclonal antibodies. J Immunol 131:232–237

Rowe M, De Gast GC, Platts-Mills TAE, Asherson GL, Webster ADB, Johnson SM (1980) Lymphocyte 5'-nucleotidase in primary hypogammaglobulinemia and cord blood. Clin Exp Immunol 39:337–343

Ruuskanen O, Pittard WB, Miller K, Pierce G, Sorensen RU, Polmar SH (1980) *Staphylococcus aureus* Cowan 1-induced immunoglobulin production in human cord blood lymphocytes. J Immunol 125:411–413

Scala G, Oppenheim JJ (1983) Antigen presentation by human monocytes: Evidence for stimulant processing and requirement for interleukin 1. J Immunol 131:1160–1166

Smith RT, Eitzman DV, Catlin ME, Wirtz OE, Miller BE (1964) The development of the immune response. Characterization of the response of the human infant and adult to immunization with Salmonella vaccines. Pediatrics 68:163–183

South MA (1972) Lack of immune response to *Hemophilus influenzae:* immune paralysis or immaturity? J Pediatr 80:348–350

Tan PLJ, Shore A (1984) Thymosin induces helper function in OKT3-positive E-rosette-negative human cord blood T cells. Scand J Immunol 20:27–34

Taylor S, Bryson Y (1985) Impaired production of γ-interferon by newborn cells in vitro is due to a functionally immature macrophage. J Immunol 134:1493–1497

Thomas RM, Linch DC (1983) Identification of lymphocyte subsets in the newborn using a variety of monoclonal antibodies. Arch Dis Child 58:34–38

Tosato G, Magrath IT, Koski IR, Dooley NJ, Blaese RM (1980) B cell differentiation and immunoregulatory T cell function in human cord blood lymphocytes. J Clin Invest 66:383–388

Uhr JW, Dancis J, Franklin EC, Finkelstein MS, Lewis EW (1962) The antibody response to bacteriophage 0174 in newborn premature infants. J Clin Invest 41:1509–1513

Vahlquist B (1949) Response of infants to diphtheria immunization. Lancet I:16–18

Van Tol MJD, Ziljsha J, Thomas CMG, Zegers BJM, Ballieux RE (1984a) Distinct role of neonatal and adult monocytes in the regulation of the in vitro antigen-induced plaque-forming cell response in man. J Immunol 134:1902–1908

Van Tol MJD, Ziljsha J, Thomas CMG, Zegers BJM, Ballieux RE (1984b) Antigen-induced plaque-forming cell responses in cultures of peripheral blood mononuclear cells of human neonates and infants. J Pediatr 105:738–744

Wakasugi N, Virelizier JL (1985) Defective IFN-γ production in the human neonate. I. Dysregulation rather than intrinsic abnormality. J Immunol 134:167–171

Wakasugi N, Virelizier JL, Arenzana-Seiscledos F, Rothhut B, Huerta JMM, Russo-Marie F, Fiers W (1985) Defective IFN-γ production in the human neonate. II. Role of increased sensitivity to the suppressive effects of prostaglandin E. J Immunol 134:172–176

Yachie A, Miyawaki T, Nagaoki T, Yokoi T, Mukai M, Uwadana N, Taniguchi N (1981) Regulation of B cell differentiation by T cell subsets defined with monoclonal OKT4 and OKT8 antibodies in human cord blood. J Immunol 127:1314–1317

Yarchoan R, Nelson D (1983) A study of the functional capabilities of human neonatal lymphocytes for in vitro specific antibody production. J Immunol 131:1222–1228

Zlabinger GJ, Mannhalter JW, Eibl MM (1983) Cord blood macrophages present bacterial antigen *(Escherichia coli)* to paternal T cells. Clin Immunol Immunopathol 28:405–411

The Ontogeny of the Immune Response: The Role of Maternal Factors

L. Å. Hanson[1], I. Adlerberth[1], B. Carlsson[1], U. Dahlgren[1], Fehmida Jalil[5], Shaukat R. Khan[4], Shakila Zaman[5], P. Larsson[2], L. Mellander[3], A. Kaleem Sheikh[4], T. Söderström[1], and A. E. S. Wold[1]

Depts. of [1]Clinical Immunology, [2]Clinical Bacteriology and [3]Pediatrics, University of Göteborg, Göteborg, Sweden;
Depts. of [4]Pediatrics and [5]Social and Preventive Pediatrics, King Edward Medical College, Lahore, Pakistan

1 Introduction

The newborn infant must be capable of coping with the many microorganisms to which it is exposed after birth. Most of the contact between the neonate and the microorganisms occurs on mucous membranes, which represent extensive surfaces of thin lining protecting the integrity of the host. For the understanding of the host-parasite relationship in the young infant and the ontogeny of immunity, we need to determine what factors microorganisms might use to colonize and possibly invade the mucosal membranes.

2 Bacterial Colonization of the Neonate

The newborn is rapidly colonized on its mucous membranes in the respiratory and gastrointestinal tract after birth (Bettelheim et al. 1976). *Escherichia coli* of various serotypes can soon be found in the intestine. In healthy Pakistani infants delivered in hospital, *E. coli* could be isolated from the stool as early as the 2nd day of life on average (mean) whereas Swedish infants also delivered in hospital had *E.coli* in the stool from the 3rd day (mean) (Table 1).

We determined the adhesins of the *E.coli* strains to try to understand how they managed to colonize the intestine. In the Swedish infants we noted that just 28% had type-1 pili, known to bind to mucus (Hanson et al. 1983). Such adhesins were found in 51% of the Pakistani infants (Table 2).

Table 1. First appearance of fecal *Escherichia coli* in Swedish and Pakistani infants

Infants			First appearance of *E.coli*	
Origin	Delivery mode	No	Mean (days)	Range
Sweden	Normal	18	2.9	1-7
Pakistan	Normal	19	1.4	1-4
Pakistan	C.section	17	2.0	1-5

Immunology of the Neonate
Burgio, Hanson, Ugazio (Eds.)

Table 2. Fecal *Escherichia coli* isolates with various hemagglutination patterns in newly delivered Swedish and Pakistani infants

Infants			Number of infants with fecal *E.coli* isolates showing:			
Origin	Delivery mode	No.	MSHA	MR(p-)HA	MR(p+)HA	only HA(−)
Sweden	Normal	18	5(28%)	7(39%)	4(22%)	5(28%)
Pakistan	Normal	19	10(53%)	4(21%)	5(26%)	6(32%)
Pakistan	C.section	17	10(49%)	0	7(41%)	5(29%)

HA, hemagglutination; *MSHA*, mannose sensitive *HA* of guinea pig red blood cells *(RBC); MR(p−)HA*, mannose resistant HA of human *RBC*, no agglutination of *p RBC; MR(p+)HA*, mannose resistant HA of human *RBC*, agglutination of *p RBC*

Most notable was, however, that 22% of the Swedish and 33% of the Pakistani infants were colonized with *E.coli* carrying mannose-resistant adhesins, including the colonization factors I, II, or other adhesins, which have been associated with capacity of the bacteria to produce enterotoxin.

In a previous study we noted that only about 17% of Swedish neonates picked up the *E.coli* from the mother at delivery. The remainder presumably received such bacteria from the hospital milieux. The mothers were simultaneously or later colonized with the same strains as their infants, if they were so at all (Gothefors et al. 1976).

3 Passive Immunity in the Neonate

3.1 Human Milk

The provision of large amounts of SIgA antibodies via the breast milk is a defense factor that can support the mucosal defense of the newborn - if breast feeding is initiated directly after birth. In many countries, the onset of breast feeding is delayed 1–3 days and microbially contaminated materials are often given instead (Hanson et al. 1986). The early colostrum may provide significant protection in spite of the small volumes produced, owing to the very high concentrations of SIgA (Carlsson et al. 1976).

The milk contains a number of other defense factors, in addition to SIgA (Hanson and Brandtzaeg 1986). Among these are oligosaccharides and/or glycolipids-glycoproteins which may have similar or identical structure as receptors for various bacteria on mucosal epithelial cells (Andersson et al. 1985). Such receptor analogues in human milk may prevent adherence of bacterial pathogens to the mucosal membranes of the breast-fed infant. We have in fact demonstrated that milk can prevent the adherence of otogenic pneumococci and *Haemophilus influenzae* to pharyngeal cells via non antibody-containing fractions, presumably holding receptor analogues (Andersson et al. 1985, 1986). The activity against pneumococci resided primarily in a low molecular weight fraction, whereas the factor inhibiting *H.influenzae* was of high molecular weight. The receptor analogues, together with the milk SIgA antibodies, may prevent infection by hindering the bacteria from reaching the mucous membranes where they might initiate infection.

The various defense factors of human milk cannot, however, eliminate bacteria from the stool. In the stool of breast-fed infants we have found *E.coli* carrying O and K antigens against which their mother's milk had SIgA antibodies (Carlsson et al. 1976, 1982). It was remarkable that the mothers had antibodies against a large variety of *E.coli,* not only those present in their intestine at the time of sampling (Gothefors et al. 1976). This could be due to a continuous and more varying exposure to *E.coli* serotypes than we have been aware of. Alternatively, long-term SIgA responses are somehow favoured in the mammary gland.

3.2 Immunoglobulin G

It is important to determine how the neonate protects itself from the colonizing microorganisms, some of which must be regarded as potential pathogens. The transplacental transfer of maternal IgG antibodies to the fetus and the role of these antibodies for protection against tissue infection is well known (Brambell 1970). Using monoclonal IgM and IgG antibodies against *E.coli* virulence factors, such as O antigen, capsules and pili, it was possible to show that given parenterally to newborn rats they could protect against bacteremia caused by perorally given *E.coli* (Söderström et al. 1983; Pluschke and Achtman 1985). These observations suggest that transferred maternal IgG antibodies may support the defense of the newborn also in the mucous membranes, helping to retain the intestinal bacteria in the gut lumen. The maternal IgG antibodies are obviously important for protection before the infant's own IgG and secretory IgA (SIgA) antibody production has started.

The maternal IgG antibodies may also depress the immune response of the offspring. In recent studies it has been shown that feeding pregnant rabbits or rats with a protein results in tolerance against this antigen in the offspring (Fig. 1) (Peri and Rothberg 1981; Dahlgren et al. 1986). The maternal antibody response consisted of IgG antibodies transferred via milk and serum, but there was also a transfer of antigen via these fluids which could have been instrumental in the tolerization of the pups (Dahlgren et al. 1986). The mechanisms for the development of tolerance in the pups to later immunization is still unknown.

4 Active Immunity in the Neonate

4.1 Immunoglobulin M

IgM antibodies are quickly and efficiently produced after birth. This is seen in the rapidly increasing serum IgM levels (Berg and Johansson 1969). In recent studies we noticed that secretory component(SC)-carrying IgM antibodies (SIgM) against *E.coli* O antigens seemed to appear regularly in the saliva of

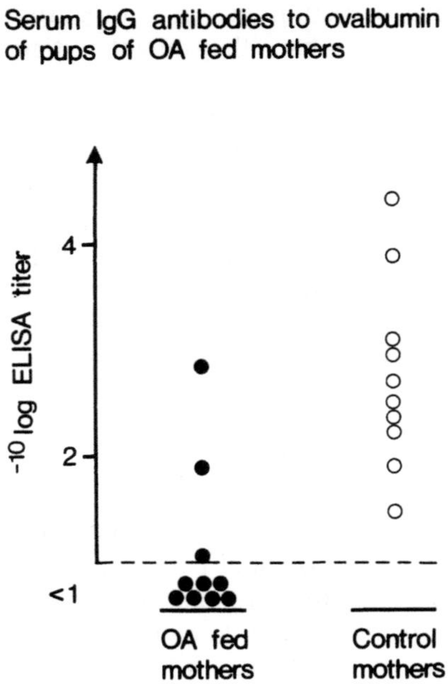

Fig. 1. IgG anti-ovalbumin *(OA)* antibodies in serum of rat pups of OA-fed mothers compared to pups of control mothers fed ordinary chow. The OA-fed rat mothers were given OA-substituted pellets during pregnancy and lactation. The pups were immunized at 2 month of age with 1 mg OA in Freund's incomplete adjuvant and serum samples were taken 21 days later. Each *dot* represents one rat pup. The antibody titers were determined with ELISA and expressed as the negative log dilution giving 0.1 absorbance unit above background

infants during the first months of life (Mellander et al. 1986a). As SIgA increased towards adult levels, SIgM decreased (Mellander et al. 1984, 1985). Their disappearence was inversely related to the appearence of SIgA, suggesting that as increased local production and transfer of SIgA took place, less SC became available for the transport of IgM into the secretion (Fig. 2a, b) (Mellander et al. 1985). The transfer of SIgM onto mucosal membranes as long as SIgA remains low may efficiently improve the defense of the mucosal membranes in the young infant.

4.2 Serum IgA

It is well known that serum IgA antibodies appear quite slowly, but there has been a controversy as to when the mucosal SIgA antibody response can add to the defense of the newborn. It has been clear, however, that SIgA can evolve earlier than, and independently of, serum IgA (for review see Hanson et al. 1980). In a recent study we could demonstrate that Pakistani infants already had significantly higher levels of SIgA antibodies against *E.coli* O antigens in saliva than Swedish infants by 1–3 weeks of age (Figs. 2a and b). At 1 month of life they had almost reached adult levels (Mellander et al. 1985). SIgA antibodies against various *E.coli* pili appeared more slowly than secretory antibodies against the O antigens (Hanson et al. 1983).

The Pakistani infants started to produce salivary SIgA antibodies against poliovirus as they were exposed, starting at 1 month of age and reaching adult levels of salivary SIgA antibodies as early as about 6 months of age (Carlsson et

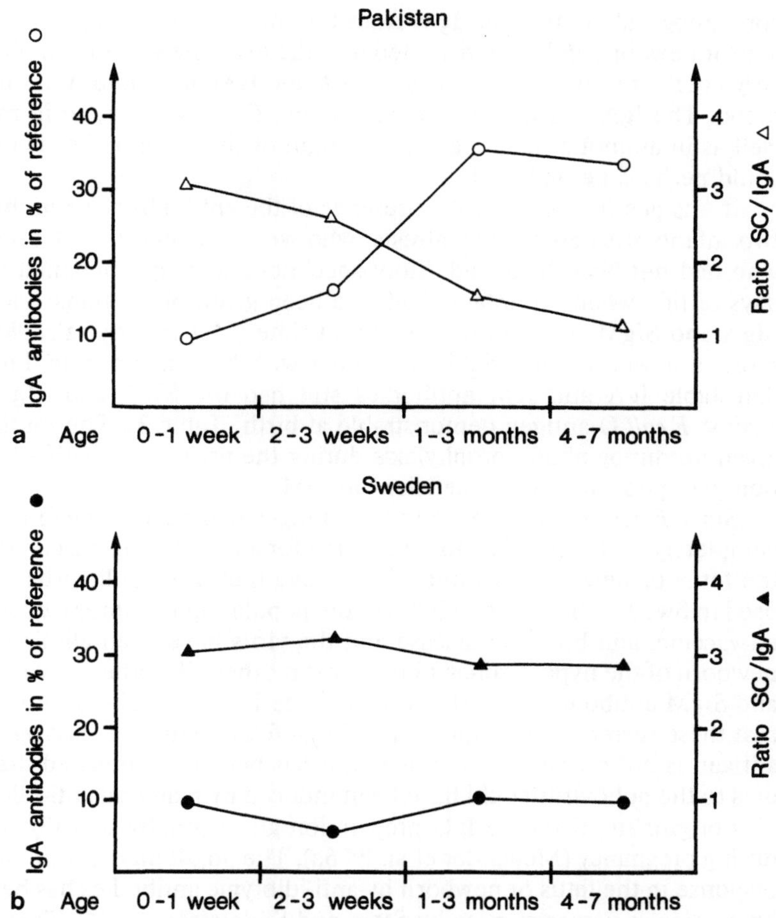

Figs. 2a, b. Secretory IgA antibodies against *E.coli* O antigens in saliva from young infants from **a** Pakistan (○) and **b** Sweden (●) determined with ELISA and expressed as percent ages of a reference. The ratio between secretory component carrying antibodies and IgA antibodies against the *E.coli* O antigens goes down in the Pakistani infants (△), inversely to the increase of the secretory IgA antibody level. In the Swedish infants the ratio (▲) remains high, since there is no increase in the level of the secretory IgA antibodies

al. 1985). SIgA antibodies against poliovirus did not appear in the saliva of Swedish children until after 3–4 injections of inactivated polio virus vaccine, since there are no wild or vaccine virus strains in the community. The Pakistani infants responded with salivary SIgA antibodies to live and inactivated vaccines as well, but it was not clear whether or not the exposure to wild poliovirus interfered with these responses.

It was remarkable that SIgA and SIgM antibodies against *E.coli* O antigens were found in the saliva already at birth (Mellander et al. 1985). The origin of these antibodies is not obvious. SIgA and SIgM are not known to pass from the mother to the infant. On the other hand, it should be considered that J-chain-containing IgA dimers and IgM from the mother could bind to SC in the fetus and/or newborn and be transported into the exocrine secretions. If this system is very efficient it may explain why no IgA and IgM antibodies were found in cord blood. The IgA and IgM antibodies against *E.coli* were found in meconium, as well as in amniotic fluid. Again, the origin of these antibodies is unknown, but could perhaps be maternal.

It was possible to show the presence of the antibodies also in the saliva from two infants with esophageal atresia, who were operated the first day of life and who had not been breastfed. Stool specimens were not obtained until about 7 days of life, when the infants had also been given blood transfusion. They had SIgA and SIgM coproantibodies at that time (Mellander et al. 1986a). Furthermore, a newborn delivered by a mother with hypogammaglobulinemia lacking detectable IgA and IgM antibodies still had the SIgA and SIgM antibodies against *E.coli* O antigen demonstrable at birth (Table 3). The mother had been given immunoglobulin prophylaxis during the pregnancy, with an IgG preparation very poor in IgA and deficient in IgM.

Since *E.coli* antigens are abundant, antigen exposure to the fetus could not be completely excluded. We chose to search for antibodies against an antigen which the fetus or newborn was unlikely to have met. Live poliovirus vaccine is not used in Sweden, but close to 100% of the population is immunized with inactivated vaccine, and both the disease and the virus have been eliminated. Still, the newborn of the hypogammaglobulinemic mother, like other newborns, had SIgA and SIgM antibodies against poliovirus type-1 antigen. The observations suggest that these secretory antibodies could be of fetal origin – at least partly. Since the antigen is not present in the milieux, it has been contemplated that the antibodies to the poliovirus could have been induced by maternal anti-idiotypic antibodies originating from the IgG preparation given prophylactically to the mother during pregnancy (Mellander et al. 1986a). The possibility of priming an immune response in the fetus or newborn by anti-idiotypic antibodies has been suggested by studies in newborn mice by Stein and Söderström (1984). They showed that monoclonal anti-idiotypic antibodies against the *E.coli* carbohydrate capsule K13 primed for protective vaccination against lethal infection with *E.coli* K13. Such protection could not be attained in vaccinated neonatal animals not primed with the anti-idiotype.

If the assumption is correct that anti-idiotypic antibodies from the mother can prime/immunize the fetus, it may have significant consequences. Newborns may in fact already be primed at birth against many of the microbial and other

Table 3. Antibodies to *E.coli* O antigens in an infant of a mother with hypogammaglobulinemia

Antibody class	IgG	IgA	IgM	SC
Saliva day 1	ND	4	19	52
Meconium	0	4	12	traces

Antibody levels were determined by ELISA and are expressed as percentages of a reference.
ND, not done because of insufficient volume

antigens to which the mother has an immune response. Vaccination in infants may not induce just primary responses, and the newborn may be better fit to respond to the many antigens it meets after birth than previously realized.

An additional aspect of the fetal immune response might be discussed. Holmberg et al. (1984) have shown that the B-cell repertoire of fetal mice contains germ-line sequences which are internally complementary, providing the possibility for a network of idiotypic interaction. Corley (1985) has suggested that autoreactive T lymphocytes can stimulate fetal B cells efficiently in conjunction with the formation of auto-anti-idiotypic antibodies, which could substitute for antigen in that sequence of events. As a result, B cells would be stimulated and proliferate and often mutate. However, only B cells remaining recognized by auto-anti-idiotypes would be further stimulated – unless maternal anti-idiotypes fit the antigen-specific receptor.

5 Summary

The newborn is exposed to a wide array of microorganisms after birth and is colonized on mucous membranes by several potential pathogens. The maternal milk contains not only large amounts of SIgA antibodies against the microorganisms present in the mother's milieus, but also receptor analogues. Such analogues, often of carbohydrate nature, may prevent microbial adherence by competition with the mucosal epithelial receptors.

The maternal IgG reaching the fetus via the placenta is certainly important in protection, and so are the IgM antibodies rapidly produced by the infant after birth. Such IgM is also transferred into secretions via the secretory component. This transport mechanism is taken over by IgA as its production in mucosal sites increases. The infant can produce significant amounts of SIgA against bacterial and viral antigens already during the first few weeks of life.

Secretory IgA and IgM antibodies are found at birth. Transfer of such antibodies from the mother has not been substantiated, but is not excluded. Our data suggest, however, that production of these antibodies may take place in the fetus, and that one inducing factor might be anti-idiotypic antibodies from the mother.

Acknowledgements. We thank Helena Kahu, Maria Badenfors, and Eva Nisshagen for excellent technical assistance. The studies were funded by the Swedish Agency for Research Cooperation with Developing Countries, the Swedish Medical Research Council (No 215), the Ellen, Walter and Lennart Hesselmann Foundation, the Swedish Insurance Companies' Fund for Medical Research and the Swedish National Bank Fund. We also thank the staff and mothers at Lady Willington Hospital, Lahore, Pakistan, who willingly cooperated in the study.

References

Andersson B, Porras O, Hanson LÅ, Svanborg Edén C, Leffler H (1985) Nonantibody containing fractions of breast mild inhibit epithelial attachment of *Streptococcus pneumoniae* and *Haemophilus influenzae*. Lancet 1:643

Andersson B, Porras O, Hanson LÅ, Lagergård T, Svanborg Edén C (1986) Inhibition of attachment of *Streptococcus pneumoniae* and *Haemophilus influenzae* by human milk and receptor oligosaccharides. J Infect Dis 153:232–237

Berg T, Johansson SGO (1969) Immunoglobulin levels during childhood with special regard to IgE. Acta Paediatr Scand 58:513–524

Bettelheim KA, Breadon A, Faiers MC, O'Farrell SM, Shootek RA (1974) The origin of O-serotypes of *Escherichia coli* in babies after normal delivery. J Hyg (Camb) 72:67–78

Brambell FWR (1971) The transmission of passive immunity from mother to young. North-Holland, Amsterdam

Carlsson B, Ahlstedt S, Hanson LÅ, Lidin-Janson B, Lindblad BS, Sultana R (1976) *Escherichia coli* O antibody content in milk from healthy Swedish mothers and mothers from a very low socio-economic group of a developing country. Acta Paediatr Scand 65:413–417

Carlsson B, Gothefors L, Ahlstedt S, Hanson LÅ, Winberg J (1976) Studies of *Escherichia coli* O antigen specific antibodies in human milk, maternal serum and cord blood. Acta Paediatr Scand 65:216–224

Carlsson B, Kaijser B, Ahlstedt S, Gothefors L, Hanson LÅ (1982) Antibodies against *Escherichia coli* capsular (K) antigens in human milk and serum – their relation to the *E.coli* gut flora of the mother and neonate. Acta Paediatr Scand 71:313–318

Carlsson B, Zaman S, Mellander L, Jalil F, Hanson LÅ (1985) Secretory and serum immunoglobulin class-specific antibodies to poliovirus after vaccination. J Infect Dis 152:1238–1244

Corley RB (1985) Somatic diversification of B cells: A role for autoreactive T lymphocytes. Immunology Today 6:178–180

Dahlgren UIH, Wold AES, Ahlstedt S, Hanson LÅ (1986) Tolerization of rat pups to ovalbumin in mother's diet. (to be published)

Gothefors L, Carlsson B, Ahlstedt S, Hanson LÅ, Winberg J (1976) Influence of maternal gut flora and colostral and cord serum antibodies on presence of *Escherichia coli* in faeces of the newborn infant. Acta Paediatr Scand 65:225–232

Hanson LÅ, Carlsson B, Dahlgren U, Mellander L, Svanborg Edén C (1980) The secretory IgA system in the neonatal period. In: Perinatal infections. Excerpta Medica/Elsevier/North-Holland, Amsterdam, pp 187–204 (Ciba Foundation symposium no. 77)

Hanson LÅ, Söderström T, Brinton C, Carlsson B, Larsson P, Mellander L, Svanborg Edén C (1983) Neonatal colonization with *Escherichia coli* and the ontogeny of the antibody response. Progr Allergy 33:40–52

Hanson LÅ, Brandtzaeg P (1986) Mucosal defense. In: Stiehm ER (ed) Immunologic disorders in infants and children, 3rd edn. Saunders Philadelphia (in press)

Hanson LÅ, Jalil F, Khan SR, Hasan R, Karlberg J, Mellander L, Carlsson B, Adlerberth I, Söderström T, Kaleen A, Lindblad BS (1986) Breast feeding in reality. In: Maternal – environmental factors in human lactation. Oaxaca, Mexico (in press)

Holmberg D, Forsgren S, Ivars F, Coutinho A (1984) Reactions amongst IgM antibodies derived from normal, neonatal mice. Eur J Immunol 14:435–441

Mellander L, Carlsson B, Hanson LÅ (1984) Appearance of secretory IgM and IgA antibodies to *Escherichia coli* in saliva during early infancy and childhood. J Pediatr 104:564–568

Mellander L, Carlsson B, Jalil F, Söderström T, Hanson LÅ (1985) Secretory IgA antibody response against *Escherichia coli* antigens in relation to exposure. J Pediatr 107:433–536

Mellander L, Carlsson B, Hanson LÅ (1986) Secretory IgA and IgM antibodies to *E.coli* and poliovirus type I antigens occur in amniotic fluid, meconium and saliva from newborns. Clin Exp Immunol (in press)

Peri BA, Rothberg RM (1981) Specific suppression of antibody production in young rabbit kits after maternal ingestion of bovine serum albumin. J Immunol 127:2520–2525

Pluschke G, Achtman M (1985) Antibodies to O-antigen of lipopolysaccharide are protective against neonatal infection with *Escherichia coli* K1. Infect Immun 49:365–370

Stein K, Söderström T (1984) Neonatal administration of idiotype or anti-idiotype primes for protection against *Escherichia coli* K13 infection in mice. J Exp Med 160:1001–1011

Söderström T, Stein K, Brinton JR CC, Hosea S, Burch C, Hansson HA, Karpas A, Schneerson R, Sutton A, Vann WF, Hanson LÅ (1983) Serological and functional properties of monoclonal antibodies to *Escherichia coli* type 1 pilus and capsular antigens. Progr Allergy 33:259–274

Mononuclear Phagocyte Function in the Perinatal Period

A. Biondi[1], G. Poli[1], C. Parravicini[2], A. Molinari[3], and A. Mantovani[1]

[1] Istituto di Ricerche Farmacologiche "Mario Negri", Milano, Italy
[2] Cattedra di Anatomia e Istologia Patologica V, Università di Milano, Italy
[3] Clinica Ostetrica e Ginecologica, Università di Milano, Osp. L. Sacco, Italy

1 Introduction

The monocyte-macrophage lineage represents a primitive first line of defense, maintained and refined in evolution. In higher organisms, while retaining their basic role of primary line of resistance, phagocytic cells have become integrated into the subsequently evolved mechanisms of specific immunity. Cooperation between lymphoid cells and macrophages occurs at different levels in the genesis of immune responses, and, in inflammatory sites, they play important roles in the expression of cell-mediated immune reactions. The role played by mononuclear phagocytes in resistance against infectious agents has long been recognized. Age-dependent resistance to infectious agents is a well-known phenomenon and host responses to a range of microorganisms, including bacteria, mycobacteria, yeasts, parasites, and viruses are relatively poor in the perinatal period in humans (Weston et al. 1977; Bellanti and Hurtado 1976). Considerable evidence has accumulated that selected functions of mononuclear phagocytes are defective in newborns and that this defect of the mononuclear phagocyte system plays a crucial role in increased susceptibility to some infectious agents (Pitt et al. 1977). In this review we will concisely discuss selected aspects of mononuclear phagocyte function in the perinatal period, emphasizing the selectivity of alterations. Some recent results on the immunobiology of placental and breast milk macrophages will also be mentioned, since, in addition to having attracted our attention, they have an obvious bearing on resistance in the perinatal period.

2 Membrane Antigens of Mononuclear Phagocytes

Cells of the mononuclear phagocyte system are considerably heterogeneous in terms of morphology, membrane phenotype, and functional properties. In recent years, monoclonal antibodies have been widely utilized to define the membrane phenotype and heterogeneity of mononuclear phagocytes (Todd and Schlossmann 1984). Monoclonal antibodies to membrane structures of monocytes have been instrumental in the discovery of a novel congenital disease related to the lack of leukocyte LFA-1 and MO1 (OKMI) (Dana et al. 1984; Beatty et al. 1984; Anderson et al. 1984; Arnaout et al. 1984). One of us (A. B.) has been involved in

Immunology of the Neonate
Burgio, Hanson, Ugazio (Eds.)

generating new monoclonal antibodies directed against mononuclear phagocytes obtained from different anatomical sites, these cells being clearly heterogeneous in terms of morphology and function. In the course of these studies, a monoclonal antibody (PAM1) directed against pulmonary alveolar macrophages has been generated and characterized (Biondi et al. 1984). PAM1 reacts, among "free" macrophages, exclusively with bronchoalveolar macrophages (Biondi et al. 1984). However, recent work using sensitive immunohistological techniques has shown that while PAM1 is expressed more strongly on alveolar macrophages, it can also be demonstrated on macrophages in other locations. Briefly, normal hystocytes and Kuppfer cells expressed PAM1 weakly; sinus histocytes in lymph nodes were moderately to strongly positive. No cell positivity for PAM1 has been found in the kidney, in the GI tract, or in the thymus (Kobzik et al. 1985). Outside the mononuclear phagocyte system, PAM1 was detected on endothelial lining cells of red pulp sinusoids, with no staining of other endothelial components in the spleen or in any other tissue. This observation is of interest considering the intriguing nature of the splenic sinusoidal lining cells.

Interestingly enough, when fetoplacental tissues were examined, PAM1 was detected on placental dendritic cells of the amniochorion. In addition to reacting with PAM1, these cells are HLA-DR$^+$ and did not react with the markers that are commonly used to define other well-characterized dendritic populations (T6, KIM, S-100, Rfd3) (Ralfkiaer et al. 1984). Cells with the same macrophage and phenotypic features (i.e., HLA-DR$^+$, PAM$^+$) have been detected in the umbilical cord with perivascular and subepithelial distribution. At the cord implant, these cells seemed to be in anatomical continuity with the dermal dendritic cells that displayed PAM1 positivity in the same tissue sections.

In the course of these studies, a monoclonal antibody (BMM$_1$) reactive with breast milk macrophages was also generated (Biondi et al. 1984). Macrophages accumulate in considerable amounts in milk during the first 5–7 days of lactation, to disappear thereafter. The functions played by milk macrophages have not been completely defined, but they may act via their microbicidal mechanisms to protect the mammary gland and neonate, and also as secretory cells (see below). It is of interest that HL60 cells are induced to differentiate along the monocyte-macrophage lineage by lymphokine express BMM$_1$ (Biondi et al. 1984).

Thus, the results obtained by studying the expression of already available monoclonal antibody-defined surface markers and generating new ones with specialized macrophage populations emphasize the heterogeneity of the mononuclear phagocyte system. Elucidation of the biological significance of the peculiar expression of PAM1 in cells presumably related to the monocyte-macrophage lineage in the fetoplacental unit and of BMM$_1$ in breast milk macrophages awaits the identification of the still elusive functional roles of the molecules identified by these monoclonal reagents.

Considerable attention has focused on membrane antigens expressed on monocytes from infants (Table 1). No alteration of monocyte-specific antigen OKM1 has been observed (Kelley et al. 1984). In contrast, in most studies in which expression of class II histocompatibility antigens (Ia) was examined, defective expression was found in the neonatal period both in humans and in rodents (Kelley et al. 1984; Stiehm et al. 1985; Kalayanarooj and Bryson 1984). In

Table 1. Alteration of selected macrophage function in newborn mononuclear phagocytes

Species	Function	Selected references
Mouse	Resistance to HSV	Hirsh et al., 1970
	Production of IFN	Hirsh et al., 1970
	Antigen presentation	Landahl, 1976
	Ia expression	Lu et al., 1979
	IL-1	Lu et al., 1979
Man	Response to newborn LDCF	Kretschemer et al., 1976
	Resistance to HSV	Mintz et al., 1980
	HLA-DR (by IFN-γ)	Kelley et al., 1984
	Response (anti-toxoplasma) to MAF (IFN-γ)	Wilson an Haas, 1984
	Accessory function for IFN-γ production	Taylor and Bryson, 1985

HSV, herpes simplex virus; *IFN,* interferon; *IL-1,* interleukin 1; *LDCF,* lymphocyte-derived chemotactis factor; *MAF,* macrophage-activating factor

rodents, the lower degree of expression of Ia by neonatal mononuclear phagocytes resulted in defective accessory function and antigen presentation by these cells (Lu et al 1979; Landahl 1976). According to two studies, lower release of interleukin-1 (IL-1) could also be involved in this defect (Lu et al 1979; Dinarello et al. 1981). In contrast, available information in humans indicates that newborn monocytes present antigen effectively (Hoffman et al. 1981). In one study, the defective production of interferon-γ (IFN-γ) by newborn lymphocyte cells was attributed to inadequate support by immature monocytes (Wakasugi and Virelizier 1985; Taylor and Bryson 1985), but Virelizier's group attributed this defect to increased sensitivity of lymphocytes to PGE suppression rather than to inadequate monocyte function (Wakasugi et al. 1985).

3 Secretory Function

Macrophages are secretory cells (for review see van Furth 1980), an aspect of these cells that may be crucial in inflammatory and immune reactions. Mononuclear phagocytes are well-known producers of arachidonic acid metabolites, potent regulators of a variety of cell function (Stenton and Parker 1980). In the course of a characterization of arachidonic acid metabolism in mononuclear phagocyte populations, we recently examined the production of arachidonate metabolism in human milk macrophages (Vicenzi et al. 1984). Milk macrophages produced considerable amounts of the prostanoids prostaglandin E_2 (PGE_2) and thromboxane B_2 (TxB_2). Human milk contains high concentrations of PGE_2 and TxB_2 (Lucas and Mitchell 1980; Reid et al 1980). These arachidonate metabolites produced by macrophages could play a crucial role in milk, PG-regulating gastrointestinal motility (Mukhopadhyay et al. 1975) and providing a cytoprotective effect on this mucosal surface (Robert et al. 1978, 1979). We speculated that mononuclear phagocytes, upon entering milk, are triggered by phagocytosis of debris and fat globules to secrete increased amounts of PG. According to this view, milk macrophages would provide a major source of these mediators important in the regulation and protection of the gastrointestinal tract of the newborn.

The capacity of the newborn monocytes to secrete and respond to lymphokines has been the object of considerable attention in recent years. Selected aspects of these studies are summarized in Tables 1–3. Peritoneal macrophages from newborn mice secrete lesser amounts of IFN upon challenge with virus, a defect which presumably contributes to enhanced susceptibility to virus infection (Hirsch et al. 1970). It is unclear whether this reflects an intrinsic defect of the macrophage capacity to produce IFN or the expression of defective handling of virus. Production of IFN-α by human cord blood leukocytes was similar to that of adult cells, whereas IFN-γ release by lymphocytes was low (Wakasugi and Virelizier 1985; Taylor and Bryson 1985; Bryson et al. 1980; Ray 1970; Kohl and

Table 2. Lymphokine production by human newborn MMC

Lymphokine	Production in newborns	Mechanism of defect	Selected reference
LIF	=	–	Hahn et al., 1976
Lymphotoxin	↓	?	Eife et al., 1974
MIF	↓	?	Stiehm et al., 1979
LDCF	↓,=	?	Kretschmer et al., 1976 Keller et al., 1983
IL-1	=,↓	–	Dinarello et al., 1981
IL-2	=	–	Wilson and Westall, 1984
MAF (?)	↓	?	Wilson and Haas, 1984
IFN-α	=	–	Ray, 1970; Kohl and Harmon, 1980
IFN-γ	↓	Immature mø; sensitivity to PGE_2	Bryson et al., 1980; Taylor and Bryson, 1985 Wakasugi and Virelizier, 1985

LIF, leukocyte inhibitor factor; *MIF*, migration inhibitory factor; *IL-2*, interleukin 2; *Mø*, macrophage ; *LDCF*, lymphocyte-derived chemotactic factor; *IL-1*, interleukin 1; *IFN*, interferon; *MAF*, macrophage-activating factor; PGE_2, prostaglandin E_2

Table 3. Selected macrophage functions unaffected in newborn mononuclear phagocytes

Species	Function	Selected reference
Mouse	Phagocytosis (↑)	Nakano et al., 1978
	Cytotoxicity to tumors (↑)	Ido et al., 1984
Man	IL-1	Dinarello et al., 1981 Taylor and Bryson, 1985
	Antigen presentation	Hoffman et al., 1981;
	Cytotoxicity to tumors	Mantovani et al., 1980

Harmon 1980). IL-1 released in the perinatal period has not been throughly investigated. In one study in mice defective IL-1 production was briefly mentioned (Lu et al. 1979), but others have reported normal IL-1 release in newborns (Dinarello et al. 1981).

4 Effector Function

Macrophages can kill microorganisms and transformed cells intracellularly or extracellularly via different pathways. Phagocytosis has been reported normal or increased in mononuclear phagocytes from neonatal mice (Nakano et al. 1978). Human cord blood monocytes and peritoneal macrophages from newborn mice are able to kill tumor cells extracellularly (Mantovani et al. 1980; Ido et al. 1984). Newborn and adult monocytes killed *Toxoplasma gondii* with equal efficiency, and survival and replication of this pathogen were comparable in newborn and adult macrophages (Wilson and Haas 1984). Exposure to lymphokine supernatants decreases the survival and replication of toxoplasma in macrophages, but newborn macrophages were less responsive to activation by lymphokine than adult cells. Moreover, newborn lymphoid cells cultured with mitogen generated lesser amounts of the lymphokine (IFN-γ ?) which activates macrophages against this pathogen (Wilson and Haas 1984).

Natural killer (NK) cells share some surface markers and functional properties with mononuclear phagocytes, though the lineage to which these cells belong has not been elucidated (Herberman and Holden 1978). NK activity is low in the perinatal period both in rodents and humans (Herberman and Holden 1978; Kaplan et al 1982). A selective deficiency of the NK cell subpopulation cytotoxic against herpesvirus-infected cells has been reported in neonates with severe systemic infections with this agent and otherwise normal immunity (Kohl 1983; Lopez et al. 1982).

5 Concluding Remarks

The information available, selected aspects of which have been briefly discussed here, strongly suggests that mononuclear phagocytes play an important role in determining enhanced susceptibility to infectious agents in the perinatal period. Alteration of the monocyte-macrophage system in the perinatal period appears relatively selective, only some functions of mononuclear phagocytes being profoundly affected. Among these, a lower expression of Ia antigens, with associated decreased antigen presentation, and of effector functions driven by lymphokines could play a pivotal role in enhancing susceptibility to infectious agents. The mechanisms responsible for alteration of selected functions of macrophages have been clarified only to a limited extent. Defective production of IFN-γ by newborn lymphoid cells has been consistently found in several studies. Decreased release of this lymphokine could play a central role in the alterations of functions found in neonatal monocytes. IFN-γ induces expression of Ia on macrophages and killing of pathogens, two aspects of the immunobiology of mononuclear phagocytes defective in newborns. Deficiency of endogenous IFN-γ could also contribute to decreased NK activity in the perinatal period.

Acknowledgements. This work was supported by "Comitato M.L. Verga" - Milan.

References

Anderson DC, Schmalstieg FC, Arnaout MA, Kohl S, Tosi MF, Dana N, Buffone GJ, Hughes BJ, Brinkley BR, Dickey WD, Abramson JS, Springer T, Boxer LA, Hollers JM, Smith CW (1984) Abnormalities of polymorphonuclear leukocyte function associated with a heritable deficiency of high molecular weight surface glycoproteins (GP138): Common relationship to diminished cell adherence. J Clin Invest 74:536–551

Arnaout MA, Spits H, Terhorst C, Pitt J, Todd III RF (1984) Deficiency of leukocyte surface glycoprotein (LFA-1) in two patients with MO1 deficiency. J Clin Invest 74:1291–1300

Beatty PG, Ochs HD, Harlan JM, Price TH, Rosen H, Taylor RF, Hansen JA, Klebanoff SJ (1984) Absence of a monoclonal-antibody-defined protein complex in boy with abnormal leukocyte function. Lancet 1:535–537

Bellanti JA, Hurtado RC (1976) Immunology and resistance to infection. In: Remington JS, Klein JO (eds) Infectious diseases of the fetus and newborn infant. Saunders, Philadelphia, p 33

Biondi A, Rossing TH, Bennett J, Todd III RF (1984) Surface membrane heterogeneity among human mononuclear phagocytes. J Immunol 132:1237–1243

Bryson YJ, Winter HS, Gard SE, Fischer TJ, Stiehm ER (1980) Deficiency of immune interferon by leukocytes of normal newborns. Cell Immunol 55:191–200

Dana N, Todd III RF, Pitt J, Springer TA, Arnaout MA (1984) Deficiency of a surface membrane glycoprotein (M01) in man. J Clin Invest 73:153–159

Dinarello CA, Shparber M, Kent EF Jr, Wolff SM (1981) Production of leukocytic pyrogen from phagocytes of neonates. J Infect Dis 144:337–343

Eife RF, Eife G, August CS, Kuhre WL, Staehr KJ (1974) Lymphotoxin production and blast cell transformation by cord blood lymphocytes: Dissociated lymphocyte function in newborn infants. Cell Immunol 14:435–442

Hahn T, Levin S, Handzel ZT (1976) Leukocyte migration inhibition factor (LIF) production by lymphocytes of normal children, newborns, and children with immune deficiency. Clin Exp Immunol 24:448–454

Herberman RB, Holden HT (1978) Natural cell-mediated immunity. In: Klein G, Weinhouse S (eds) Advances in cancer research, vol 27. Academic, New York, p 305

Hirsch MS, Zisman B, Allison AC (1970) Macrophages and age-dependent resistance to herpes simplex virus in mice. J Immunol 104:1160–1165

Hoffman AA, Hayward AR, Kurnick JT, Defreitas EC, McGregor J, Harbeck R (1981) Presentation of antigen by human newborn monocytes to maternal tetanus toxoid-specific T-cell blasts. J Clin Immunol 1:4–8

Ido M, Uno K, Inaba K, Aotsuka Y, Muramatsu S (1984) Ontogeny of "macrophage" function: IV. Newborn mouse macrophages strongly suppress tumour cell growth and readily acquire cytolytic activity in comparison with adult macrophages. Immunology 52:307–317

Kalayanarooj S, Bryson Y (1984) Decreased Ia AG on human neonatal macrophages; a marker of maturation? Clin Res 32:258A

Kaplan J, Shope TC, Bollinger RO, Smith J (1982) Human newborns are deficient in natural killer activity. J Clin Immunol 2:350–355

Keller MA, Kidd RM, Leake RD, Everett SL (1983) Lymphocyte-derived chemotactic factor production by neonatal lymphocytes. Pediatr Res 17:799–802

Kelley VE, Fiers W, Strom TB (1984) Cloned human interferon-γ but not interferon-β or -α, induces expression of HLA-DR determinants by fetal monocytes and myeloid leukemic cell lines. J Immunol 132:240–245

Kobzik L, Godleski JJ, Biondi A, O'Hara C, Todd RF (1985) Immunohistologic analysis of a human pulmonary alveolar macrophage antigen. Clin Immunol Immunopathol 37:217–219
Kohl S (1983) Defective infant antiviral cytotoxicity to herpes simplex virus-infected cells. J Pediatr 102:885–888
Kohl S, Harmon MW (1980) Human neonatal leukocyte interferon production and natural killer cytotoxicity in response to herpes simplex virus. J Interferon Res 3:461–463
Kretschmer RR, Stewardson P, Gotoff SP (1976) Immunologic disorders in children. IMJ 149:52–57
Landahl CA (1976) Ontogeny of adherent cells. I. Distribution and ontogeny of A cells participating in the response to sheep erythrocytes in vitro. Eur J Immunol 6:130–134
Lopez C, Kirkpatrick D, Fitzgerald P (1982) The role of NK (HSV-1) effector cells in resistance to herpesvirus infections in man. In: Herberman RD (ed) NK cells and other natural effector cells. Academic, New York, p 1445
Lu CY, Calamai EG, Unanue ER (1979) A defect in the antigen-presenting function of macrophages from neonatal mice. Nature 282:327–329
Lucas A, Mitchell MD (1980) Prostaglandins in human milk. Arch Dis Child 55:950–952
Mantovani A, Bar Shavit Z, Peri G, Polentarutti N, Bordignon C, Sessa C, Mangioni C (1980) Natural cytotoxicity on tumour cells of human macrophages obtained from diverse anatomical sites. Clin Exp Immunol 39:776–784
Mintz L, Drew WL, Hoo R, Finley TN (1980) Age-dependent resistance of human alveolar macrophages to herpes simplex virus. Infect Immun 28:417–420
Mukhopadhyay A, Rattan S, Goyal RK (1975) Effect of prostaglandin E_2 on esophageal motility in man. J Appl Physiol 39:479–481
Nakano K, Hosokawa T, Muramatsu S (1978) Ontogeny of macrophage function: I. Phagocytic activity and A-cell activity of newborn and adult mouse peritoneal macrophages. Dev Comp Immunol 2:505–518
Pitt J, Barlow B, Heird WC (1977) Protection against experimental necrotizing enterocolitis by maternal milk: I. Role of milk leukocytes. Pediatr Res 11:906–909
Ralfkiaer E, Stein H, Plesner T, Hou-Jensen K, Mason D (1984) In situ immunological characterization of Langerhans cells with monoclonal antibodies: Comparison with other dendritic cells in skin and lymph nodes. Virchows Arch [A] 403:401–412
Ray CG (1970) The ontogeny of interferon production by human leukocytes. J Pediatr 76:94–98
Reid B, Smith H, Friedman Z (1980) Prostaglandins in human milk. Pediatrics 66:870–872
Robert A, Lancaster C, Nezamis JE, Hanchar AJ (1978) Cytoprotective prostaglandins exogenous or endogenous can maintain gastric secretory function. Gastroenterology 74:1086
Robert A, Nezamis JE, Lancaster C, Hanchar AJ (1979) Cytoprotection by prostaglandins in rats. Prevention of gastric necrosis produced by alcohol, HCl, NaOH, hypertonic NaCl, and thermal injury. Gastroenterology 77:433–443
Stenton WF, Parker CW (1980) Prostaglandins, macrophages, and immunity. J Immunol 125:1–5
Stiehm ER, Winter HS, Bryson YJ (1979) Cellular (T-cell) immunity in the human newborn. Pediatrics 64 (5/2 [Suppl]): 814–821
Stiehm ER, Sztein MB, Steeg PS, Mann D, Newland C, Blaese M, Oppenheim JJ (1985) Deficient DR antigen expression on human cord blood monocytes. Reversal with lymphokines. Clin Immunol Immunopathol (in press)
Taylor S, Bryson YJ (1985) Impaired production of γ-interferon by newborn cells in vitro is due to a functionally immature macrophage. J Immunol 134:1493–1497
Todd RF III, Schlossmann SF (1984) Utilization of monoclonal antibodies in the characterization of monocyte-macrophage differentiation antigens. In: Bellanti JA, Herscowitz HB (eds.) The reticuloendothelial system: A comprehensive treatise, vol 6: Immunology. Plenum, New York, p 87
Van Furth R (ed) (1980) Mononuclear phagocytes. Functional aspects. Nijhoff, The Hague
Vicenzi, E, Biondi A, Bordignon C, Rambaldi A, Donati MB, Mantovani A (1984) Human mononuclear phagocytes from different anatomical sites differ in their capacity to metabolize arachidonic acid. Clin Exp Immunol 57:385–392
Wakasugi N, Virelizier J-L (1985) Defective IFN-γ production in the human neonate. I. Dysregulation rather than intrinsic abnormality. J Immunol 134:167–171
Wakasugi N, Virelizier J-L, Arenzana-Seisdedos F, Rothhut B, Mencia Huerta J-M, Russo-Marie F, Fiers W (1985) Defective IFN-γ production in the human neonate: II. Role of increased sensitivity to the suppressive effects of prostaglandin E. J Immunol 134:172–176

Weston WL, Carson BS, Barkin RM, Slater GD, Dustin RD, Hecht SK (1977) Monocyte-macrophage function in the newborn. Am J Dis Child 131:1241-1242
Wilson CB, Haas JE (1984) Cellular defenses against *Toxoplasma gondii* in newborns. J Clin Invest 73:1606-1616
Wilson CB, Westall J (1984) Interleukin 2 (IL2) and macrophage activation factor (MAF) production by newborn lymphocytes. Clin. Res. 32:267A

Mechanisms of Abnormal Neutrophil Function in the Human Neonate: Prospects for Therapy

HARRY R. HILL and FULVIO SACCHI

Departments of Pathology and Pediatrics, University of Utah College of Medicine, Salt Lake City, UT 84132, USA; and the Departments of Pediatrics and Chemotherapy, University of Pavia, Pavia, Italy

1 Infections in the Neonate Characteristic of Neutrophil Disorders

Human neonates have an increased incidence of infections when compared with older children and adults (Hill 1985; Santos and Hill 1982; Klein and Marcy 1983). The infections suffered by term and especially premature infants are usually due to bacterial and fungal agents that may not be particularly virulent in older individuals unless they are immunocompromised. These include pathogens such as group B streptococci, *Staphylococcus epidermidis, Listeria monocytogenes, Escherichia coli* and *Candida albicans* (Santos and Hill 1982; Klein and Marcy 1983). Furthermore, newborns often develop infections in peripheral sites, such as on the skin, in subcutaneous tissues, or in the lungs. Phagocytic cells, including polymorphonuclear leukocytes and macrophages, are critical to the defense of these areas. Miles et al. (1957) have shown, in experimental animals, that phagocytic cells must arrive at a site of microbial invasion, ingest the pathogen, and kill it within a critical 2-4 h period. If this does not occur, the animal will eventually develop a larger local or systemic infection.

Human infants are especially prone to develop group B streptococcal (GBS) infection of the lung and often of the skin (Hill 1984a). Such infections rapidly gain access to the bloodstream and become systemic in nature. While antibody and complement are quite important in host defense against GBS, adequately functioning phagocytic cells are essential for the killing of these organisms. Adults with normally functioning PMNs seldom suffer GBS infection in spite of the fact that most have low to absent levels of antibody to GBS. In contrast, adult diabetic patients, who also have PMN abnormalities, have a high incidence of skin and urinary tract infections due to these organisms (Hill 1984a).

Cutaneous abscesses at the site of monitor probe insertion and mucocutaneous candida infections are also exceedingly common in human neonates (Santos and Hill 1982). While lymphocytes certainly play a role in immunity to candida, the neutrophil is a critical aspect of the host defense system against such organisms in the newborn period before delayed-type hypersensitivity has developed. Thus, these types of infections also tend to point to abnormal neutrophil function as a major developmental host defense abnormality in the fetus and newborn.

Immunology of the Neonate
Burgio, Hanson, Ugazio (Eds.)

2 Defects in PMN Function in Human Neonates

In order to adequately contain invasion by pathogenic microbes, PMNs must first sense the presence of inflammatory mediators released either directly from the organism or from the complement system. These factors serve to prime or activate the cell making it more likely to adhere, aggregate, chemotax, and generate toxic oxygen molecules (Hill et al. 1974; Barbour et al. 1980). After adherence to the vascular endothelial cells nearest the site of tissue invasion, the cells aggregate, move through the vessel wall and then chemotax, or move in a unidirectional fashion toward the invading microbe. If the particle has been opsonized by antibody and/or complement, the phagocytic cell attaches to it and may internalize it. Attachment and ingestion result in the activation of a respiratory burst in PMNs and macrophages and the generation of a number of microbicidal oxygen products including superoxide, hydrogen peroxide, hydroxyl radical, and hypochlorite ion (Hill 1984b).

The human neonate possesses PMNs that demonstrate a number of functional abnormalities. First, the cells will aggregate in response to chemotactic stimuli, such as C5a or the formyl peptide, formylmethionyleucylphenylalanine (FMLP). This response is not reversible, however, as it is in PMNs of adults (Mease et al. 1981). Second, the chemotactic response of PMNs from human neonates is depressed compared with that of PMNs from adults (Miller 1975, 1979; Klein et al. 1977). Once the phagocytes are exposed to an opsonized particle, however, phagocytic uptake appears to be intact. Respiratory burst activity as measured by the generation of chemiluminescence or nitroblue tetrazolium dye reduction appears to be normal or even increased in human neonatal PMNs (Shigeoka et al. 1979, 1981). Phagocytosis and actual bacterial killing also appear to be normal in term neonates when sufficient opsonin is supplied (Dossett 1969). Under certain stressful conditions, however, subtle abnormalities in respiratory burst activity and intracellular killing have been observed. These conditions include instances where the amount of opsonin is limited (Miller 1969) in premature infants (Wright et al. 1975) and in stressed infants with bacterial infections (Shigeoka et al. 1979, 1981). Whether this subtle abnormality in respiratory burst activity under conditions of stress is of importance clinically remains to be determined.

3 Events Associated with PMN Activation and Function

A complex series of events results in PMN activation and function. First, chemotactic factors bind to specific receptors on the surface of the cell (Williams et al. 1977). This triggers a series of events which include changes in membrane potential (Seligman et al. 1980), ion fluxes (Naccache et al. 1977), a transient, early increase in cyclic adenosine monophosphate levels (Simchowitz et al. 1980; Jackowski and Sha'afi 1979), the appearance of microtubules and actin polymerization (Rao and Varani 1981). Abnormalities in any of these critical processes could result in profound functional abnormalities, such as those observed in the PMNs of human neonates.

4 Abnormalities in the Activation of PMNs From Human Neonates

In an attempt to understand the mechanisms involved in the movement abnormality in the PMNs of human neonates, we have examined some of the critical events in the activation sequence described above. First, we have studied chemotactic factor-receptor binding in the PMNs from neonates and compared the results with those obtained with PMNs from adults. Receptor binding was studied employing tritium-labeled FMLP (New England Nuclear, Boston, Mass) as previously described (Sacchi and Hill 1984; Anderson et al. 1981). Specific binding was determined by subtracting total binding from that when an excess of unlabeled FMLP was added. Experiments have now been carried out at 37° C, 4° C, and 0° C, and we have found no difference in total or specific FMLP binding to the PMNs of neonates when compared with cells from adults (Sacchi and Hill 1984; Hill, Augustine and Sacchi, submitted for publication). Thus, at least under the conditions of our assay, chemotactic factor receptor binding appears to be normal with the PMNs of neonates. Similar results have been reported by Anderson and coworkers (1981) and by Strauss and Snyder (1984) at 37° C or 4° C.

We have recently examined chemotactic-factor-stimulated membrane potential changes by the PMNs of neonates (Sacchi and Hill 1984). In order to examine this critical activation event, we have employed a spectrofluorometric technique and the membrane potential sensitive dye 3,3-dipentyloxacarbocyanine [$DiOC_5(3)$]. This dye partitions itself according to the potential of the cell membrane. A decrease in fluorescence emission is associated with depolarization, while an increase probably results from hyperpolarization of the membrane (Seligman and Gallin 1980). As can be seen in Fig. 1, exposure of PMNs from adults to 2.5×10^{-8} *M* FMLP caused an increase in fluorescence emission. This most probably follows a brief depolarization of the membrane, which Seligman and Gallin (1980) have associated with respiratory burst activity. The observed

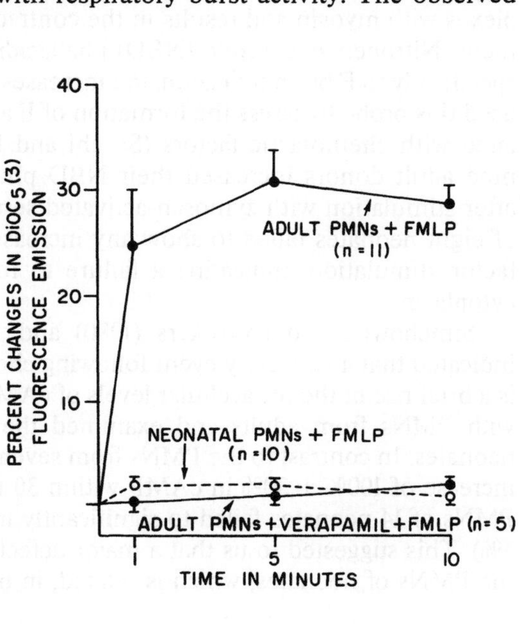

Fig. 1. Membrane potential changes induced in PMN from 11 adults and 10 neonates by exposure to 2.5×10^{-8}mM FMLP as determined by $DiOC_5$ (3) fluorescence emission. (From Sacchi and Hill 1984)

increase in $DiOC_5$ (3) fluorescences is most probably the result of a hyperpolarization caused by the extrusion of calcium by a calcium-activated ATP-dependent pump (Lagast et al. 1984). In contrast to the response by the PMNs of adults, cells obtained from neonates fail to show any significant increase in fluorescence (Fig. 1).

Because calcium is felt to play a major role in cellular activation and to drive the ATP-dependent calcium pump, we next determined the effects of chemotactic factors on intracellular free calcium levels. In these studies, we employed the Ca^{2+}-sensitive dye Quin 2/AM (Sacchi and Hill 1984). As can be seen in Fig. 2, there was a prompt rise in Quin 2/AM fluorescence after chemotactic factor stimulation of PMNs from adults, which indicates a marked increase in the level of intracellular free calcium. This occurred in calcium-containing media (1.6 m*M*) and in calcium-deprived media to which 2 m*M* EGTA had been added. The increase in free calcium observed following chemotactic factor stimulation is, therefore, most likely derived from intracellular calium stores rather than the external environment. Verapamil, a calcium channel-blocking agent which has also been reported to block intracellular release of calcium, completely blocked the rise in free calcium observed in chemotactic-factor-stimulated PMNs from adults.

Neutrophils from neonates also showed an increase in free calcium after FMLP stimulation (Fig. 2), but the response was approximately one-third less than that of the adult cells in calcium-containing or calcium-depleted medium. We speculate that the intracellular stores of calcium may be decreased in the PMNs of neonates, and that the lower levels of free calcium reached may not be sufficient to stimulate the calcium-activated ATP-dependent calcium pump.

We next examined a later event in the process leading to cell movement. Upon chemotactic factor stimulation, filamentous or F actin is formed from the polymerization of monomeric G actin (Rao and Varani 1981). The F actin complexes with myosin and results in the contractile forces necessary for cell movement. Nitrobenzoxadiazole (NBD) phallacidin is a fluorescent probe that binds specifically to F but not G actin and increases its fluorescence emission. We have used this probe to assess the formation of F actin in PMNs from neonates stimulated with chemotactic factors (Sacchi and Hill, submitted). Neutrophils from nine adult donors increased their NBD phallacidin fluorescence by 45%–55% after stimulation with zymosan-activated serum. In marked contrast, the PMNs of eight neonates failed to show any increase in fluorescence after chemotactic factor stimulation, indicating a failure to form filamentous actin within their cytoplasm.

Simchowitz and coworkers (1980) and Jackowski and Sha'afi (1979) have indicated that a very early event following chemotactic factor-receptor interaction is a brief rise in the intracellular levels of cAMP. We have confirmed these results with PMNs from adults and examined the responses of PMNs from human neonates. In contrast to the PMNs from seven adults, which had a prompt average increase of 100% ± 30% in cAMP within 30 to 60 s, after FMLP stimulation the PMNs of 14 neonates failed to significantly increase their cAMP content (10% ± 5%). This suggested to us that a major defect in signal transduction is present in the PMNs of neonates, which is related, in part, to cAMP generation.

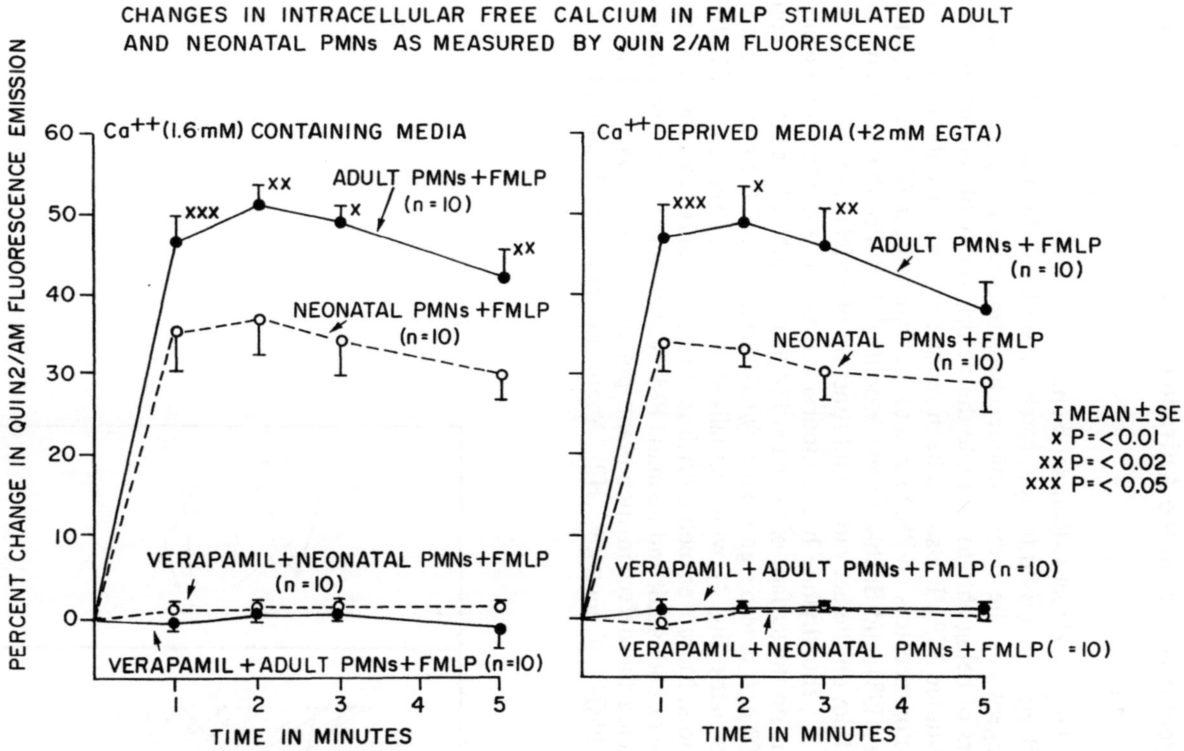

Fig.2. Changes in intracellular free calcium in FMLP-stimulated PMN from ten adults and ten neonates as determined by Quinn 2/AM fluorescence emission. (From Sacchi and Hill, 1984)

5 Effects of Pentoxifylline on the Activation and Function of PMNs From Neonates

Pentoxifylline (Trental, Hoechst-Roussel, Sommerville, NJ) is a methylxanthine derivative which has been used to treat cerebrovascular and peripheral vascular disease. This agent has the rather unique characteristic among phosphodiesterase inhibitors of being able to increase the deformability of cell membranes (Marcel 1979; Muller 1979). It also has been reported to increase the number of chemotactic factor receptors on PMNs and to increase cellular ATP levels (Sullivan and Mandell 1984; and Buchanan and Moodley 1977). For these reasons, we examined its effect on the activation and function of PMNs from neonates. As shown in Fig. 3, pentoxifylline had a modest but significant effect to increase FMLP receptors on the surface of neonatal PMNs ($p=0.01$). The drug, in a concentration of 100 μg/ml, also significantly ($p<0.001$) improved the chemotaxis of PMNs from neonates (Fig. 4), while actually decreasing the response of PMNs from adult donors. Moreover, pentoxifylline treatment markedly enhanced the cAMP response of both adult and neonatal PMNs to chemotactic factor stimulation and corrected the defect in membrane potential change that we had observed in the PMNs of neonates (Hill, Augustine and Sacchi, submitted for publication).

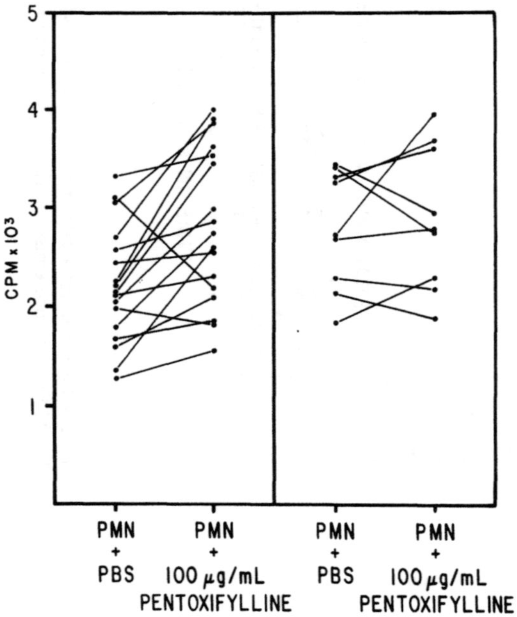

Fig. 3. Effect of pentoxifylline (100 μg/ml) on specific ^{3}H-FMLP binding to adult and neonatal PMNs

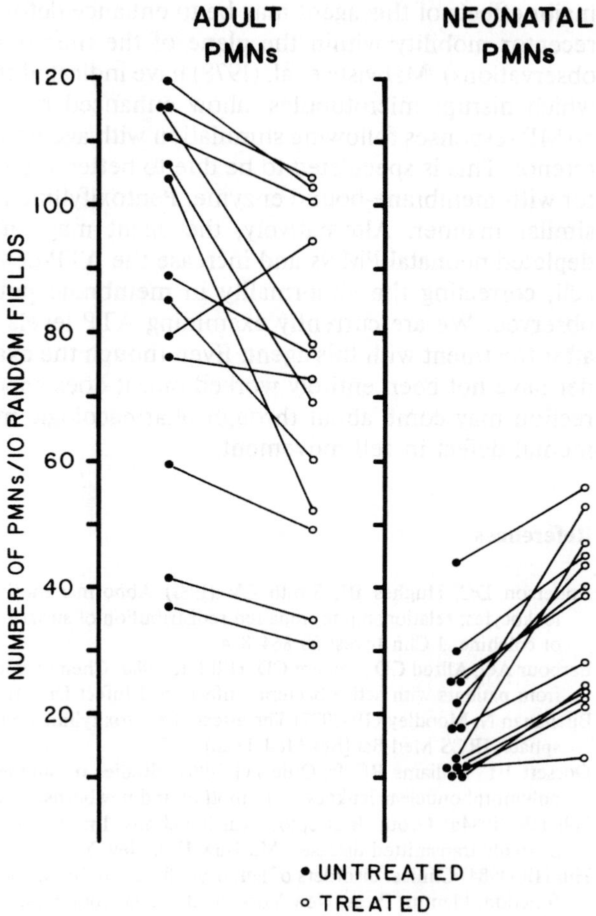

Fig. 4. Effect of pentoxifylline (100 μg/ml) on the chemotactic response of PMNs from adults and neonates

6 Discussion

The PMNs of premature and term neonates are clearly defective in their ability to move toward a chemotactic stimulus. The reasons for this abnormality are not clearly understood. Miller et al. (1975), Kimura et al. (1981), and Anderson et al. (1981) have shown that PMNs of neonates have rigid cytoskeletal structures that are incapable of deforming normally and redistributing adhesion sites or surface receptors following chemotactic factor stimulation. We have now also shown that even though these cells appear to interact normally with chemotactic factors, they fail to undergo critical changes in membrane potential, do not increase their cAMP levels, and do not polymerize gel actin to filamentous actin within the cytoplasm. It appears, therefore, that the chemotactic signal is not being adequately transduced from the surface to internal enzymes and structures. The effect of pentoxifylline to correct many of these abnormalities was not unexpected since this agent has been reported to increase cell deformability, increase cAMP responses, and elevate ATP levels within the cell. Exactly how it accomplishes these effects is not completely understood. We suspect that one of the major effects of this agent may be to enhance deformability and allow enhanced receptor mobility within the plane of the membrane (Hill et al., unpublished observations). Malwista et al. (1978) have indicated that agents such as colchicine, which disrupt microtubules, allow enhanced receptor mobility and increased cAMP responses following stimulation with agents such as histamine and isoproterenol. This is speculated to be due to better approximation of occupied receptor with membrane-bound enzyme. Pentoxifylline may be exerting its effect in a similar manner. Alternatively, the agent may enhance ATP levels in ATP-depleted neonatal PMNs and increase the ATP-dependent calcium pump in the cell, correcting the abnormality in membrane potential change that we have observed. We are currently examining ATP levels in neonatal cells before and after treatment with this agent. Even though the exact mechanisms of this disorder have not been entirely worked out, it does appear at least possible that correction may come about through pharmacologic manipulation of this developmental defect in cell movement.

References

Anderson DC, Hughes BJ, Smith CV (1981) Abnormal motility of neonatal polymorphonuclear leukocytes: relationship to impaired redistribution of surface adhesion sites by chemotactic factor or colchine. J Clin Invest 68:864–874

Barbour AG, Allred CD, Solberg CD, Hill HR (1980) Chemiluminescence production by granulocytes from patients with active bacterial infection. J Infect Dis 141:14–26

Buchanan N, Moodley GP (1977) The effect of pentoxifylline on human erythrocyte adenosine triphosphate. IRCS Med Sci [Bio Med Tech] 5:43

Dossett JH, Williams RC Jr, Quie PG (1969) Studies on interaction of bacteria, serum factors and polymorphonuclear leukocytes in mother and newborns. Pediatrics 44:49–57

Hill HR (1984a) Group B streptococcal infections. In: Holmes KK, Mordh PA, Sparling PF (eds) Sexually transmitted diseases. McGraw Hill, New York, pp 397–407

Hill HR (1984) Clinical disorders of leukocyte function. In: Snyderman R (ed) Regulation of leukocyte function. Plenum Press, New York, pp 345–393 (contemporary topics in immunobiology)

Hill HR (1985) Diagnosis and treatment of sepsis in the neonate. In: Root RK, and Sande MA (eds) Septic shock: contemporary issues in infectious diseases. Churchill Livingstone, New York, p 291–232

Hill HR, Gerrard J, Hogan NA, Quie PG (1974) Hyperactivity of neutrophil leukotactic responses during active bacterial infection. J Clin Invest 53:996–1002

Jackowski S, Sha'afi (1979) Response of adenosine cyclic 3',5'-monophosphate levels in rabbit neutrophils to the chemotactic peptide formyl-methionyl-leucyl-phenylalanine. Mol Pharmacol 16:473–481

Kimura GM, Miller ME, Leake RD, Raghunathan R, Cheung ATW (1981) Reduced concanavalin A capping of neonatal polymorphonuclear leukocytes (PMNs). Pediatr Res 15:1217–1273

Klein JO, Marcy SM (1983) Bacterial sepsis and meningitis. In: Remington JS and Klein JO (eds) Infectious diseases of the fetus and newborn infant. Saunders, Philadelphia, p 679

Klein RB, Fischer TJ, Gard SE, Biberstein M, Rich KC, Stiehm ER (1977) Decreased mononuclear and polymorphonuclear chemotaxis in human newborns, infants and young children. Pediatrics 60:467–472

Lagast H, Lew PD, Waldvogel FA (1984) Adenosine triphosphate dependent calcium pump in the plasma membrane of guinea pig and human neutrophils. J Clin Invest 73:107–115

Malawista SE, Oliver JM, Rudolph SA (1978) Microtubules and cyclic AMP in human leukocytes: On the order of things. J Cell Biol 77:881–886

Marcel GA (1979) Red cell deformability: Physiological, clinical and pharmacologic aspects. J Med 10:409–416

Mease AD, Burgess DP, Thomas PJ (1981) Irreversible neutrophil aggregation: a mechanism of decreased newborn neutrophil chemotactic response. Am J Pathol 104:98-102

Miles AS, Miles EM, Burke J (1957) The value and duration of defense reactions of the skin to primary lodgement of bacteria. Br J Exp Pathol 38:79-96

Miller ME (1969) Phagocytosis in the newborn infant: Humoral and cellular factors. J Pediatr 74:255-259

Miller ME (1975) Developmental maturation of human neutrophil motility and its relationship to membrane deformability. In: Bellanti JA, Dayton DH (eds) The phagocytic cell in host resistance. Raven, New York, pp 295-307

Miller ME (1979) Phagocyte function in the neonate: selected aspects. Pediatrics 64:709-712

Muller R (1979) Pentoxifylline - a biomedical profile. J Med 10:307-329

Naccache PH, Showell J, Becker EL, Sha'afi RI (1977) Transport of sodium, potassium and calcium across rabbit polymorphonuclear leukocytes: effect of chemotactic factors. J Cell Biol 73:428-444

Rao KMK, Varani J (1981) Actin polymerization induced by chemotactic peptide and concanavalin A in rat neutrophils. J Immunol 129:1605-1607

Sacchi F, Hill HR (1984) Defective membrane potential changes in neutrophils from human neonates. J Exp Med 160:1247-1252

Santos JI, Hill HR (1982) Bacterial infection of the neonate. In: Wedgwood RJ, Davis SD, Ray CG, Kelley VC (eds) Infections in children. Harper and Row, Philadelphia, pp 179-202

Seligman BE, Gallin EK, Martin DL, Shain W, Gallin JI (1980) Use of lipophilic probes of membrane potential to assess human neutrophil activation. J Clin Invest 66:493-503

Seligman BE, Gallin EK, Martin DL, Shain W, Gallin JI (1980) Interaction of chemotactic factors with human polymorphonuclear leukocytes: studies using a membrane potential-sensitive cyanine dye. J Membr Biol 52:257-262

Shigeoka AO, Charette RP, Wyman ML, Hill HR (1981) Defective oxidative metabolic responses of neutrophils from stressed neonates. J Pediatr 98:392-398

Shigeoka AO, Santos JI, Hill HR (1979) Functional analysis of neutrophil granulocytes from healthy, infected and stressed neonates. J Pediatr 95:454-460

Simchowitz L, Fischbein LC, Spilberg I, Atkinson JP (1980) Induction of a transient elevation in intracellular levels of adenosine 3',5'-cyclic monophosphate by chemotactic factor: an early event in human neutrophil activation. J Immunol 124:1482-1491

Strauss RG, Snyder EL (1984) Chemotactic peptide binding by intact neutrophils from human neonates. Pediatr Res 18:63-66

Sullivan GW, Mandell GL (1984) Pentoxifylline increases chemotactic receptors and enhances chemotaxis of human neutrophils. Clin Res 32:517A

Williams LT, Snyderman R, Pike MC, Lefkowitz RJ (1977) Specific receptor sites for chemotactic peptides on human polymorphonuclear leukocytes. Proc Natl Acad Sci 74:1204-1208

Wright WC Jr., Ank BJ, Herbert J, Stiehm ER (1975) Decreased bactericidal activity of leukocytes of stressed newborn infants. Pediatrics 56:569-584

Complement Function in the Neonate

F. Sessions Cole

The Division of Cell Biology, The Children's Hospital, 300 Longwood Avenue, Boston, MA 02115, USA; and The Joint Program in Neonatology, The Brigham and Women's Hospital, 75 Francis Street, Boston, MA 02115, USA

1 Complement Cascade Activation

The complement cascade consists of 18 plasma proteins which upon activation regulate the inflammatory response, effect bacteriolysis, solubilize immune complexes, and opsonize microorganisms. Activation of the complement cascade may occur via the antibody-dependent classical pathway (Cl, C2, and C4) or the antibody-independent alternative pathway (factor B and factor D) (Arnaout 1985; Pangburn 1983; Reid 1981). Activation of early acting components of both pathways occurs by limited proteolysis. Complexing of a single dimer of IgG or IgM with antigen on cell surfaces activates a single Cl molecule (Borsos 1965). Two subcomponents of Cl, Clr and Cls, are sequentially activated, and Cls activates C4 and C2 by limited proteolytic cleavage. An unstable enzyme complex is formed between cleavage fragments of C4 and C2, C4b2a, which activates C3 by cleavage of a 9-kD fragment from the alpha chain of C3. A C3-converting enzyme which cleaves C3 at the same site can also be generated via the alternative pathway (Pangburn 1983). Factor B is a peptide structurally and functionally similar to C2. It is cleaved after complexing with C3b by factor D, a serine protease analogous to Cls. The alternative pathway C3 convertase (C3bBb) is a complex of the 70-kD activation product of factor B, Bb, and C3b. Because the larger activation product of C3, C3b, is part of this complex, a positive amplification loop is established. Activation of C3 by either pathway results in sequential activation of the terminal components C5–C9 and in deposition of C3b, the major activation product of C3, on cell surfaces. Upon activation of the terminal components, assembly of a macromolecular membrane channel by protein-protein interaction occurs, and membrane integrity of the cell or microorganism is interrupted. Deposition of C3b on cell surfaces provides an opsonin for polymorphonuclear and mononuclear leukocytes. Finally, complement activation results in liberation of small peptides from C3 (C3a), C4 (C4a), and C5 (C5a) with important proinflammatory and immunoregulatory activities (Hugli 1981). These anaphylatoxins, 8–10 kD peptides cleaved from the alpha chains of C3, C4, and C5, during activation, enhance vascular permeabilily, increase vascular smooth muscle tone, and provide stimuli for directed migration of polymorphonuclear leukocytes. Availability of early acting complement components in tissues and at sites of inflammation is critical for initiation and maintenance of an inflammatory response. When the content of transplacentally acquired IgG and classical

Immunology of the Neonate
Burgio, Hanson, Ugazio (Eds.)

pathway proteins are limited, the newborn infant depends upon alternative pathway proteins as effectors of host defense against bacteria. To understand the mechanisms which regulate the availability of complement proteins, investigators have studied the ontogeny of serum complement proteins and functional opsonic activity of neonatal sera.

2 Ontogeny of Serum Complement Proteins

Early studies in fetal sheep and piglets suggested that functional activities of individual serum complement components were low or undetectable early in gestation and rose in the third trimester to 50%–66% of adult concentrations (Colten 1968b; Day 1969, 1970; Geiger 1972a, b). Hepatic synthesis of individual components in vitro also increased during the third trimester. Because complement proteins studied to date in animals are not transplacentally transported, several investigators speculated that the increases in serum concentrations of these proteins result from increases in hepatic synthesis rates. Similar studies have been performed in the human. Both C3 and Cl INH are detectable in human fetal sera as early as 5–6 weeks of gestation (Gitlin 1969). By 26–28 weeks, both C3 and Cl INH concentrations rise from 1% of adult levels (C3) and 20% of adult levels (Cl INH) to approximately 66% of adult concentrations. Multiple investigators have demonstrated that in cord blood of infants of various gestational ages, both immunochemically and hemolytically detectable complement proteins are found in lower concentrations than in corresponding maternal sera or pooled adult sera (Adamkin 1978; Adinolfi 1970, 1975; Ballow 1974; Kohler 1968; Mills 1979; Stevenson 1980; Stossel 1973; Strunk 1979). On the basis of studies of genetically determined, structurally distinct complement variants in paired maternal-cord sera, no transplacental passage from mother to infant of C3, C4, factor B, or C6 has been observed (Johnston 1979; Propp 1968; Teisberg 1971). In addition, the presence of detectable amounts of C2 and Cl INH in cord blood but not maternal sera in women with genetic deficiencies of these proteins suggests that fetal-maternal transport of complement proteins does not occur (Alper 1976).

To investigate a possible mechanism to account for increasing serum concentrations in the human fetus, Gitlin and Biasucci (1969) examined synthesis by fetal liver of 15 plasma proteins, including C3 and Cl INH, using biosynthetic labelling in tissue culture followed by radioimmunoelectrophoresis. They detected C3 and Cl INH synthesis by fetal liver as early as 4–5 weeks of gestation and in all 15 fetal livers examined through 18 weeks of gestation. While other fetal tissues synthesized individual complement components (e.g., Cl by segments of large and small intestine of a 19-week fetus) (Colten 1968a), the liver was the primary source of complement proteins examined (Adinolfi 1968).

In vitro synthesis of other complement proteins, including C2, C4, C5, and factor H, by human fetal liver has been demonstrated (Adinolfi 1967, 1981; Colten 1972, 1973). A marked increase in C4 synthesis by fetal liver occurs at approximately 15 weeks gestation coincident with a rise in serum concentration. As demonstrated in animal studies, changes in fetal serum concentrations of complement proteins result from increases in net hepatic synthesis rates during gestation. The functional importance of lower concentrations of alternative pathway proteins for the newborn infant was suggested by studies which examined opsonic activity in neonatal sera.

3 Importance of Complement Function in the Newborn Infant

3.1 Deficiency of Serum Complement Opsonic Activity

In a study of opsonic activity of cord blood, Stossel et al. (1973) showed that 6 of 40 cord sera examined had less than 66% of the opsonic activity of pooled adult serum. No correlation was detected between C3 and IgG concentrations. However, each of these six sera had lower factor B concentrations than other newborn or adult sera. Other investigators have found similar results (Mills 1979). These findings suggest that lowered serum concentrations of alternative pathway components may limit opsonic activity and thereby contribute significantly to enhanced susceptibility of the newborn infant to systemic infection. Techniques were not available when these studies were performed to investigate the possibility that synthesis of complement proteins at extrahepatic sites might also limit complement availability.

While the liver is the primary source of serum complement proteins, cells of the monocyte-macrophage series provide the major extrahepatic source of complement production. Although not a detectable fraction of serum complement concentrations, locally produced complement components are critical in the initiation and maintenance of an inflammatory response prior to changes in vascular permeability which permit influx of soluble and cellular elements from blood (Ezekowitz 1984). Initial studies in the guinea pig and the human suggested that mononuclear phagocyte production of C2, C3, C4, and factor B in vitro was characteristic of the tissue from which the cells were isolated and the inflammatory state of the tissue (Cole 1982, 1983, 1985). For example, in the human, adult blood monocytes synthesize C2 and factor B at approximately equal rates, while breast milk macrophages produce 3.5-fold more C2 than factor B, and bronchoalveolar macrophages produce 9-fold more C2 than factor B. To determine whether these cells could respond to a stimulus which might be anticipated to enhance complement production, Strunk et al. (1985) incubated human monocytes and tissue macrophages in the presence of lipopolysaccharide (LPS) isolated from several gram-negative organisms by the Westphal procedure (Morrison 1979). LPS is a highly conserved substance found in the outer membrane of many gram-negative organisms (Morrison 1979). It is composed of a carbohydrate domain (the O antigen and core polysaccharide) and a lipid domain (lipid A). Lipid A consists of two lipid precursors, lipid X and lipid Y. The Westphal extraction procedure permits isolation of LPS without detectable protein contamination. Strunk et al. (1985) observed a dose-dependent 15-fold increase in the synthesis of C3 and a 3-fold increase in factor B. These changes were specific in that no changes in the synthesis of C2, lysozyme, or total protein were detected. Using a polysaccharide deficient mutant of *S. minnesota* and lipid A precursors, lipid X and lipid Y, the lipid A portion of LPS was determined to be the structural requirement for this effect. Using RNA blot analysis, the regulation of these effects was found to be pretranslational.

Development of these techniques permitted examination of the possibility that regulation of synthesis of C3 and factor B in tissues by neonatal mononuclear phagocytes might also limit availability of proteins critical for immunologic recognition and effector function. Because gram-negative organisms have accounted for 33% of systemic bacterial infections in the United States over the last 20 years (Siegel 1981), we examined the effects of LPS on monocytes isolated from cord blood (Sutton, personal communication).

3.2 Unresponsiveness of Cord Blood Monocytes to LPS

Monolayers of cord blood monocytes were established and maintained as previously described (Einstein 1976). All pregnancies were full term. Despite comparable rates of constitutive synthesis of C3 and factor B as well as C2, lysozyme, and total protein in cord blood and adult monocytes, cord blood monocyte synthesis of C3 and factor B was unresponsive to LPS. To account for this unresponsiveness, several possibilities were examined: (a) no differences between adult and cord blood monocytes in proportions of cells which were nonspecific esterase positive (greater than 95%), and which expressed the cell surface glycoprotein Mol (greater than 90%) were noted; (b) the viability of adult and cord blood monocytes was verified by trypan blue dye exclusion before and after incubation with LPS; (c) adult monocytes from males, nonpregnant females, and females immediately after delivery responded to LPS; (d) the route of delivery of the infant did not affect LPS responsiveness; (e) cord blood monocytes did not respond to LPS from three different strains of *E. coli;* (f) length of incubation of the cells with LPS (up to 48 h), time in culture prior to incubation with LPS (2–10 days), and concentrations of LPS used (150–1500 ng/ml) did not enhance LPS responsiveness. Two other LPS-inducible functions of mononuclear phagocytes were examined. LPS induced comparable increases in adult and cord blood monocytes in PMA-stimulated superoxide anion production and C3b-mediated phagocytosis. These data suggested that cord blood monocyte unresponsiveness to LPS was not due to quantitative differences in mononuclear phagocyte functions or qualitative differences in cell populations. We next investigated the possibilities that cord blood monocyte C3 and factor B synthesis might be regulated by LPS subcomponents or soluble factors produced by adult monocytes.

Incubation of cord blood monocytes in lipid X and lipid Y resulted in increases in C3 and factor B synthesis comparable to those observed in adult monocytes. When LPS isolated by the trichloroacetic acid procedure, in which endotoxin associated protein is copurified with LPS, and endotoxin-associated protein alone were used, increased C3 and factor B synthesis comparable to adult monocytes was noted. Mechanisms which might account for cord blood monocyte unresponsiveness to LPS are an inability to process LPS to its subcomponents or to recognize LPS without its associated protein. To determine whether LPS responsiveness is mediated by one or more soluble factors, two preliminary experiments were performed. First, adult and cord blood monocyte monolayers were coincubated in the presence of LPS. The cord blood monocytes had no effect on LPS-induced increases in C3 and factor B synthesis in adult monocytes, but coincubation resulted in enhanced C3 and factor B synthesis in cord blood monocytes. These data suggested that LPS-treated adult monocytes release one or more factors which increase cord blood monocyte responsiveness to LPS. Secondly, cord blood monocytes were incubated in the presence of trichloroacetic acid-extracted LPS and dexamethasone, ETYA, or indomethacin. Dexamethasone and ETYA abrogated the effects of TCA-extracted LPS on C3 and factor B synthesis by cord blood monocytes, while indomethacin had no effect. These data thus demonstrate that C3 and factor B synthesis is inducible in cord blood monocytes, and that al lipooxygenase pathway metabolite may be involved in mediating LPS responsiveness in cord blood monocytes. These studies do not exclude the possibilities that differences between adult and cord blood monocytes in cell surface recognition of LPS, intracellular processing of LPS, or factors from other cells (e.g., lymphokines) may also account for the unresponsiveness of cord blood monocyte C3 and factor B synthesis to LPS.

4 Conclusion

Although the increased susceptibility of the human newborn infant to systemic bacterial infection is multifactorial, studies of the ontogeny of serum complement concentrations, of hepatic synthesis of these proteins, and functional opsonic activity of neonatal sera have suggested that lower serum complement concentrations occur due to lower hepatic synthesis rates, and that lower serum concentrations lead to decreased capacity to activate the complement cascade. Recent techniques have permitted the study of the regulation of synthesis of individual complement proteins by cells of the mononuclear phagocyte series. These cells regulate availability of labile early acting components in tissues and at sites of inflammation. These studies demonstrate a specific difference between adult and cord blood monocytes in the capacity to respond to LPS, a highly conserved constituent of the outer membrane of gram-negative bacteria. Although adult mononuclear phagocytes increase synthesis of C3 and factor B by 15- and 3-fold, respectively, in response to LPS, cord blood monocytes are unresponsive. These data suggest that complement function in the newborn infant is regulated not only by ontogeny of hepatic synthesis but also by constitutive and inducible synthesis of early acting complement proteins in tissues by mononuclear phagocytes.

References

Adamkin D, Stitzel A, Urmson J (1978) Activity of the alternative pathway of complement in the newborn infant. J Pediatr 93:604–608
Adinolfi M, Gardner B (1967) Synthesis of beta-1-e and beta-1-c components of complement in human fetuses. Acta Paediatr Scand 56:450–457
Adinolfi M, Gardner B, Wood CBS (1968) Ontogenesis of two components of human complement: beta-1-e and beta-1-c/beta-1-a globulins. Nature 219:189–190
Adinolfi M (1970) Levels of two components of complement (C4 and C3) in human fetal and neonatal sera. Dev Med Child Neurol 12:306–313
Adinolfi M, Beck SE (1975) Human complement C7 and C9 in fetal and newborn sera. Arch Dis Child 50:562–564
Adinolfi M, Dobson NC, Bradwell AR (1981) Synthesis of two components of human complement, beta-1-H and C3bINA, during fetal life. Acta Paediatr Scand 70:705–712
Alper CA, Rosen FS (1976) Genetics of the complement system. Adv Hum Genet 7:141–188
Arnaout MA, Colten HR (1985) Complement C3 receptors, structure and function. Mol Immunol, in press
Ballow M, Fang F, Good RA (1974) Developmental aspects of complement components in the newborn. Clin Exp Immunol 18:257–270
Borsos T, Rapp HJ (1965) Complement fixation on cell surfaces by 19S and 7S antibodies. Science 150:505
Cole FS, Schneeberger EE, Lichtenberg NA, Colten HR (1982) Complement biosynthesis in human breast milk macrophages and blood monocytes. Immunology 46:429–443
Cole FS, Matthews WJ, Rossing RH, Shotola NA, Gash DJ, Pennington JE (1983) Complement biosynthesis by human bronchoalveolar macrophages. Clin Immunol Immunopathol 27:153–159
Cole FS, Auerbach HS, Goldberger G, Colten HR (1985) Tissue specific pretranslational regulation of complement production in human mononuclear phagocytes. J Immunol 134:2610–2616
Colten HR, Gordon JM, Borsos T, Rapp HJ (1968a) Synthesis of the first component of human complement in vitro. J Exp Med 128:595–616
Colten HR, Silverstein AH, Borsos T (1968b) Ontogeny of the first component of sheep complement. Immunol 15:459–461
Colten HR (1972) Ontogeny of the human complement system: in vitro biosynthesis of individual complement components by fetal tissues. J Clin Invest 51:725–750
Colten HR (1973) Biosynthesis of the fifth component of complement (C5) by human fetal tissues. Clin Immunol Immunopathol 1:346–352
Day NKB, Pickering RJ, Gewurz H (1969) Ontogenetic development of the complement system. Immunol 16:319–325
Day NKB, Gewurz H, Pickering RJ (1970) Ontogenetic development of Clq synthesis in the piglet. J Immunol 104:1316–1321
Einstein LP, Schneeberger EE, Colten HR (1976) Synthesis of the second component of complement by long-term cultures of human monocytes. J Exp Med 143:114–126
Ezekowitz RAB, Sim RB, Hill M, Gordon S (1984) Local opsonization by secreted macrophage complement components. Role of receptors for complement in uptake of zymosan. J Exp Med 159:244–253
Geiger H, Day NKB, Good RA (1972a) Ontogenetic development and synthesis of hemolytic C8 by piglet tissues. J Immunol 108:1092–1096
Geiger H, Day NKB, Good RA (1972b) Ontogenetic development of the later complement components in fetal piglets. J Immunol 108:1098–1102
Gitlin D, Biasucci A (1969) Development of gamma-G, gamma-A, gamma-M, beta-1-c/beta-1-a, C'1 esterase inhibitor, ceruloplasmin, transferrin, hemopexin, haptoglobin, fibrinogen, plasminogen, alpha-1-antitrypsin, orosomucoid, beta-lipoprotein, alpha-2-macroglobulin, and prealbumin in the human conceptus. J Clin Invest 48:1433–1446
Hugli TE (1981) The structural basis for anaphylatoxin and chemotactic functions of C3a, C4a, and C5a. CRC Crit Rev Immunol 1:321–366
Johnston RB, Alterburger KM, Atkinson AW, Curry RH (1979) Complement in the newborn infant. Pediatrics [Suppl] 64:781–786

Kohler PF (1968) Quantitative comparison of complement in the mother and newborn. Fed Proc 27:491–492

Mills EL, Bjorksten B, Quie PG (1979) Deficient alternative complement pathway activity in newborn sera. Pediatr Res 13:1341–1344

Morrison DC, Ryan JL (1979) Bacterial endotoxins and host immune response. Adv Immunol 28:293–352

Pangburn MK (1983) Activation of complement via the alternative pathway. Fed Proc 42:139–144

Propp RP, Alper CA (1968) C'3 synthesis in the human fetus and lack of transplacental passage. Science 162:672–673

Reid KBM, Porter RR (1981) The proteolytic activation system of complement. Annu Rev Biochem 50:433–464

Siegel JD, McCracken GH (1981) Sepsis neonatorum. N Engl J Med 304:642–646

Stevenson DK, Hsu YP, McMorrow M (1980) Component concentrations and activation of the complement system in neonatal illness. A preliminary study of necrotizing enterocolitis. Eur J Pediatr 134:255–259

Stossel TP, Alper CA, Rosen FS (1973) Opsonic activity in the newborn: role of properdin. Pediatr 52:134–137

Strunk RC, Fenton LJ, Gaines JA (1979) Alternative pathway of complement activation in full term and premature infants. Pediatr Res 13:641–644

Strunk RC, Whitehead AS, Cole FS (1985) Pretranslational regulation of the synthesis of the third component of complement in human mononuclear phagocytes by the lipid A portion of lipopolysaccharide. J Clin Invest (in press)

Teisberg P (1971) Genetics of the C3 system: family, mother/child, and association studies. Hum Hered 21:458–469

Modulation of Nonspecific Defense Mechanisms in the Neonate

Francesco Laurenti

Institute of Pediatric Clinic University of Rome, Viale Regina Elena 324, Rome, Italy

1 Introduction

In the neonate there is a striking biological paradox: the deficiency of defense mechanisms is associated with an elevated risk of tissue damage due to activation of the inflammatory process. In addition, a variety of defense patterns can be elicited, i.e., multifactorial immaturity of several components; latent activation of some activities; episodes of sudden, frequently uncontrolled functional hyperactivity; and exhaustion or deactivation of various metabolic responses.

This variable behaviour of neonatal defense mechanisms corresponds to different clinical features, such as the well-known elevated incidence and severity of the infections, occurrence of sudden complications (shock, disseminated intravascular coagulation [DIC], shock lung, hemorrhage), and tissue damage due to the efflux of toxic intermediates. Thus, the old concept of neonatal defense impairment can be modified in the more extensive term "dysregulation."

In the following, an attempt will be made to pick out a thread linking the divergent biological and clinical events in such a way as to allow a generally acceptable interpretation of their interconnections.

2 Heterogeneity of Defense Mechanisms and Modulation Process

The variable patterns by which the defense mechanisms take effect suggest a modulation by means of both intrinsic and extrinsic factors, the former depending on the developmental and structural performance of components, the latter on the interactions between systems and exogenous agents. Therefore, it is worthwhile focusing one's attention on some advances recently reached in this field and on the related biological-clinical implications. These aspects include: developmental immaturity and modulating factors, latent activation of inflammatory components at birth, modifications of defense responses following neonatal infections, and other disorders.

Immunology of the Neonate
Burgio, Hanson, Ugazio (Eds.)

2.1 Developmental Immaturity and Modulating Factors

A comprehensive profile indicating the level of immunological development in the human neonate has already been delineated by several authors (Miller 1978; Davies and Gothefors 1984). On reviewing the previous works, they defined the most relevant concepts concerning the multifactorial immaturity of specific and nonspecific defense. In this context, it may be helpful to review some modulating factors. First of all, it is convenient to remark that variable and quite conflicting results depend on the different methods and experimental conditions: the percentage rate of serum opsonizing factors, microbial species, particle-neutrophil ratio, and other variables affect, for instance, phagocytosis and killing. Thus, a functional overload is followed by an impaired response.

Another point concerns the pathway of stimulation: soluble (i.e., phorbol meristate acetate; PMA) or particulate stimuli (i.e., zymosan and germs) give different effects. Superoxide, for instance, is generated in adequate amounts from either neonatal or adult neutrophils after PMA challenge; in contrast, zymosan is a more rapid and powerful trigger (Shigeoka et al. 1981; Strauss and Snyder 1983). This means that different transductional pathways and common steps are involved in each specific stimulus, and the kinetic of cellular response depends on the route of activation. Physiological and pathological factors operating in living organisms probably follow the same rules.

2.2 Latent Activation at Birth

Several authors suggest a status of inapparent activation of inflammatory components at birth in connection with the physiological and pathological events of pregnancy and parturition. Data mostly concern the stimulation of respiratory burst in neonatal phagocytes: hexose monophosphate shunt (HMPS) activity is increased either basally or after stimulation (Anderson et al. 1974), the NBT dye test elicits a certain degree of spontaneous activation (Park et al. 1970; Anderson et al. 1974; Shigeoka et al. 1979), and superoxide is released at a higher rate than in adult cells (Strauss and Snyder 1983; Shigeoka et al. 1981). Ambruso et al. (1981) observed partial degranulation and lactoferrin depletion in neonatal cells, depending on birth events. Strauss and Hart (1981) remarked that both neonatal and maternal neutrophils exhibit spontaneous concanavalin A (ConA) capping, meaning oxidative burst. Finally, Frazier et al. (1982) found altered metabolic conditions in neutrophils, in relation to delivery modalities: a reduction of oxygen consumption, a defective HMPS activity, and an impaired NBT dye test were demonstrated in subjects who experienced all stages of labor, whereas the same alterations could not be found in those who chose to have cesarean sections. Therefore, labour seems to affect the functional neutrophil integrity, and the gestational variations of epinephrine, corticosteroids, fatty acids, and arachidonic acid derivatives can explain these modifications. The labour acts as primary stimulus and the subsequent hyporesponsiveness at birth follows the rules of functional deactivation (McPhail and Snyderman 1984).

The passage to a richer oxygen extrauterine environment and the onset of pulmonary respiration can themselves activate the respiratory burst of phagocytes, as they enhance the synthesis of antioxidant enzymes (Strauss et al. 1980; Frank and Groseclose 1984).

Amniotic fluid can also activate some mechanisms. Hammerschmidt et al. (1984) found C3a and C5a anaphylotoxins in samples of normal plasma incubated with amniotic fluid, thus demonstrating complement activation. The implicated factor has been recognized in a lipid-rich particulate fraction deriving from cellular debris and increasing with gestational age, as well as with the occurrence of fetal-maternal abnormalities. Amniotic fluid stimulates also platelets, as proven by specific aggregation assays (Salem et al. 1982). Laurenti et al. (1985, unpublished data) investigated the amniotic fluid ability to induce granulocyte aggregation: centrifuged samples of amniotic fluid collected during cesarean section were added to a cytochalasin B-treated granulocyte suspension in a standard aggregometer recorder system. A defined aggregation curve occurred in those cases affected by fetal or maternal abnormalities.

In summary, we can conclude that the more physiological the birth, the more preserved is the integrity of phagocyte responses. In contrast, as abnormal labour and pregnancy affect many organs and functions, similarly they alter the homoeostasis of the inflammatory process.

2.3 Neonatal Infections and Modulation of Defense Response

In contrast to adult neutrophils, neonatal phagocytes show a drop of function and metabolic responses in the course of infections. It is well known that circulating neutrophils decrease because of impaired recruitment or depletion (Christensen and Rothstein 1980), microbicidal capacity declines (Wright et al. 1975), and the NBT test - which in healthy neonates shows a certain degree of spontaneous reduction - becomes paradoxically hyporesponsive (Cocchi et al. 1971; Shigeoka et al. 1979).

However, a very short period of stimulation can not be surely excluded, considering the difficulties in intercepting the earliest phases of infection in neonates (Davies and Gothefors 1984). We monitored chemotaxis in preterm infants with infections and observed a divergent behavior: a transient increase of chemotaxis index was evident in localized infections, while a complete suppression occurred in sepsis. After recovery, values returned to basic levels in both instances (Laurenti et al. 1980).

Direct and complement-mediated endotoxin-neutrophil interactions evoke alterations in phagocytes. Endotoxin at different dilutions produces hyperadhesiveness, enzymatic release, and oxidative burst in neutrophils, followed by deactivation (Dahinden et al. 1983). The arachidonic acid derivatives seem not to be involved, as indomethacin does not inhibit the effect (Dahinden et al. 1983) in experimental models.

Independently from clinically manifest infections, Scheifele et al. (1981) suggest that isolated and transient hyperendotoxemias deriving from bacterial overgrowth in the small gut of mother or infant frequently occur in neonate without visible clinical consequences. These hyperendotoxemias might be able to activate phagocyte respiratory burst in vivo, since a few picograms of lipopolysaccharide give the same effect.

2.4 Neonatal Disorders and Defense Mechanism Modulation

Apart from infections, severe neonatal diseases, such as asphyxia, hyaline membrane disease (HMD), cerebral hemorrhage, cardiac failure, are frequently associated with defense mechanism alterations. Although the earliest reports established an impairment of defense in stressed newborns (Wright et al. 1975; Shigeoka et al. 1981), more recent studies consider the possible development of hyperactive reactions. Therefore, both an explosion and a suppression of reactivity can occur. For a better understanding of the causes of these contradictory phenomena, we need to underline two biological aspects: (a) the functional intersystem integration and (b) the cellular circuits controlling activation, amplification, and desensitization.

2.4.1 Functional Integration of Nonspecific Defense Mechanisms

Nonspecific immunity includes primary defense systems (complement, granulocytes, and monocyte macrophages) and secondary systems (prekallikrein, coagulation, fibrinolysis, and arachidonic acid cycle). The former is more directly committed to pathogen elimination, while the latter is mostly implicated in inflammatory process. A functional net closely connects the various elements at different steps so that homoeostasis can be maintained.

Each system includes an inert precursor, which oversees the normal "status" and recruits other defenses when it is activated by perturbing agents. Activating factors can be directed (i.e., Fc fraction of Ig for Cl) or unspecific, so that many cross-reactions can occur and activation can be achieved in different ways (Table 1). Each system produces several effectors and mediators diffusing like a cluster, which can defend or attack the host according to the specific situation. In fact, the multiple interaction of mediators generate different effects, such as balanced, impaired, potentiated, or amplified responses (Yurt 1984). The control systems include specific inhibitors as well as intrinsic circuits of cellular deactivation which cooperate to contain the inflammation inside the limits of the defense capacity (Table 2).

2.4.2 Kinetics of Cellular Activation and Deactivation

Recent studies emphasize the meaning of intracellular biochemical events controlling the kinetics of biological activation, amplification, and suppression, and their relevance in clinical manifestations. Indeed, activating stimuli can follow different transductional routes, as well as share common steps. Therefore, the final response depends on the activation pathway (Sklar 1984). For instance,

Table 1. Interactions between precursors and activators

Precursors	Activators
C1	Fc fragment of Ig; C reactive protein; plasmin; trypsin; Hageman factor
C3	Properdin; B-D factor; endotoxin; kallikrein; elastase
Hageman Factor	Endotoxin; surfaces; collagen; plasmin, trypsin; activated factor XI; kallikrein (positive feedback mechanism)
Prekallikrein	Activated Hageman factor
Plasminogen	Activated Hageman factor; kallikrein
Arachidonic acid	Phospholipase A_2-C; free radicals
Neutrophils	Particles; surfaces; chemotactic factors; endotoxin; interleukin I; leukotrienes
Macrophages	Particles; chemotactic factors; C3a; antigens; activated T cells
Platelets	Collagen; ADP; trombin; thromboxane
Mast cells	Kinins; endotoxin; chemotactic factors; cationic proteins

particulate ligands and chemoattractants (zymosan, C5a, formyl methionine-leucine-phenylalanine - fMLP) are receptor mediated, employ arachidonate or cyclic nucleotides as messengers, need Ca^{++} and show a rapid activation followed by an immediate inhibition; in contrast, soluble stimuli, such as PMA, directly trigger protein kinase C and elicit a slower, more prolonged effect (Fig. 1). In addition, sequential stimulation studies demonstrate that the same cells can develop either an amplified (priming effect) or a suppressive response (deactivation). Priming effect generally occurs when a neutrophil is preexposed to a certain stimulus and then to a different challenge; the suppressive response

Table 2. Control mechanisms: Cross-reactive connections between inhibitors and targets

Inhibitors or Inactivators	Targets
Alphal-proteinase inhibitor	Kallikrein; trypsin; plasmin; thrombin; elastase; (H_2O_2 myeloperoxidase halide system)
Alpha-macroglobulin	Kallikrein; trypsin; plasmin; thrombin
Antithrombin III	Kallikrein; plasmin; factors X - XI; neutrophil proteases
Alpha 2-antiplasmin	Kallikrein; plasmin
C_1esterase Inhibitor	C_1esterase; plasmin; kallikrein; Hageman factor
β1-Inhibitor	C3b-Bb
C3bINA	C3b
Kininases I-II	Kinins; (C5a des-arginine)
Carboxy-peptidases	C3a - C5a

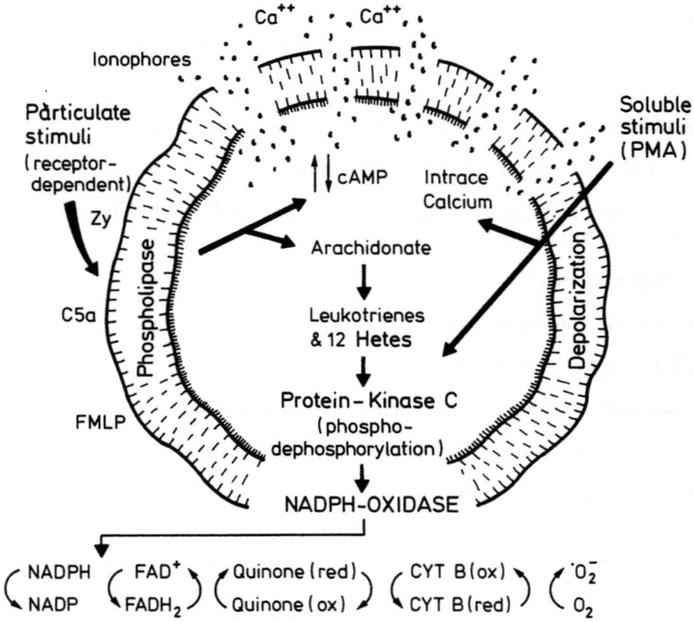

Fig. 1. Pathways of NADPH oxidase activation and intracellular events in neutrophil

is more easily elicited after homologous subsequent stimulation (McPhail and Snyderman 1984; Goldstein 1984; Sklar 1984).

These experiments give an idea of the possible pathways drawn in living organisms: the multitude of factors operating in sick neonates probably mimics the same variety of routes, achieving different degrees of response. Although the ultimate mechanisms are unknown, clinical observations suggest an intensive activation, deactivation, and amplification work of inflammatory cells in the neonate.

2.4.3 Nonspecific Activation of Defense Mechanisms in the Neonate

It is well known to neonatologists that stressed neonates can develop DIC, shock, and other disorders. Although these diseases are multifactorial, the involvement of nonspecific defense mechanisms undoubtedly plays a role. The basic events are still obscure and need further investigation. However, an important point is the occurrence of hematochemical alterations, such as hypoxemia, acidosis, ischemia, ionic and metabolic imbalances, which usually develop in stressed infants and operate as perturbing stimuli of cellular membranes. The main site is the vascular endothelium and the microcirculation, which constitutes the universal stroma for a myriad of reactions. From injured cells enzymes, proteases, peptides, and other active split products are released and trigger the previously inert precursors of the inflammatory branches (C1, prekallikrein, Hageman Factor [HF], phospholipases, NADPH oxidase) (Yurt 1984). As additional factors, it is worthwhile noting the anatomical impairment of term and preterm neonatal structures, and the unstable homeostatic balance.

The routes of activation are intricate (Carmona et al. 1984). Collagen, lipopolysaccharides, and tissue enzymes trigger the HF-dependent pathway; from the assembled HFā-prekallikrein-high-molecular-weight kininogen, kallikrein frees kinins and plasmin. The complement cascade is triggered in several ways: by tissue proteases, by kallikrein which splits Clr and Cls, and by plasmin which cleaves Cls. NADPH oxidase of phagocytes can be perturbed by arachidonate derivatives, fatty acids, and split products from the complement and clotting systems (Carmona et al. 1984). Alterations of pH, Ca^{++}, and oncotic pressure alter the membrane fluidity disclosing a net of channels, through which ionic fluxes move, whereas acidosis and hypothermia enhance the respiratory burst.

An important point concerns the proposed pathway through which hypoxia and ischemia can activate defense mechanisms, considering the frequent occurrence of these alterations in neonatal disorders. Demopoulos et al. (1980) suggest that the reduced PaO_2 unbalances the components of the chain committed to electron transport. In this connection, coenzyme Q (CoQ) dissociates functionally from the remaining intermediates and reacts with the spare oxygen assuming a radical configuration (oxygen penetrates the inner mitochondrial membrane at 4–7 mmHg pressure because of its high solubility in nonpolar lipoidal structures). In this way, the chain of the free radical generation is triggered, followed by activation of the inflammatory cells: lipoperoxidation of endothelial cells reduces the continuous prostacyclin (PgI_2) synthesis; the vasoconstrictive and cell-aggregating activity of thromboxane and leukotrienes prevails; platelets and granulocytes aggregate, perpetuating the cycle of toxic oxygen species, proteases and other substances. McCord (1985) proposes an alternative free radical generation pathway, in connection with the postischemic reperfusion phase. Hypoxia reduces the ATP synthesis by promoting the anaerobic metabolism. The following reduction of cellular energy allows Ca^{++} to penetrate the cytosol and activate a Ca^{++}-dependent protein kinase, converting xanthine dehydrogenase (present in normally perfused tissues) to xanthine oxidase, which represents a well-known free radical generating system (Fig. 2).

3 Clinical Implications

Biological modulation of defense mechanisms is followed by a wide range of clinical manifestations. The primary aspect concerns the basic impairment of the immunologic defense against the pathogens, which leads to the well known increased neonatal mortality in infections. The remaining aspects are related to the variable pathways of activation, from which different clinical consequences rise.

Latent activation of the inflammatory process does not usually display detectable clinical signs, and this condition can be recognized only with the help of laboratory assays. For instance, Sangstad et al. (1982) demonstrated the activation of the prekallikrein-kinin system in neonates with HMD, and Laurenti et al.

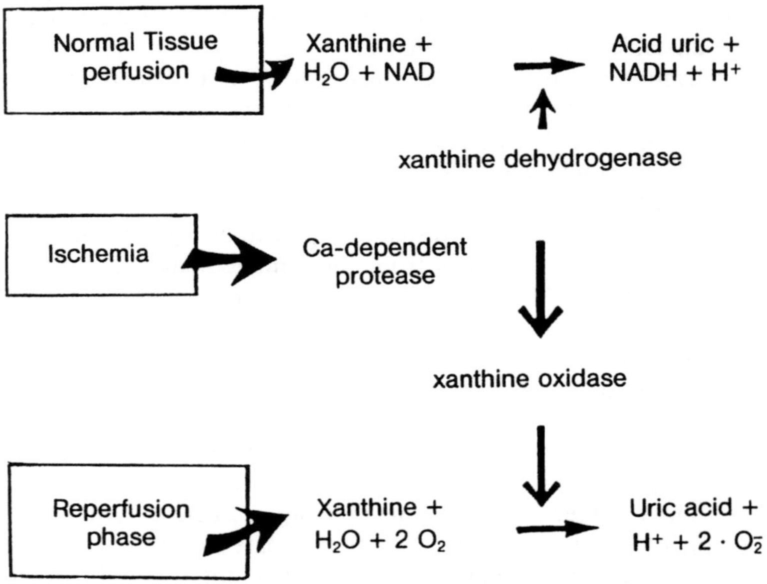

Fig. 2. Generation of free radicals in postischemic reperfusion phase

(unpublished data) extended the same observation to neonates showing different severe disorders such as HMD, asphyxia, and cerebral haemorrhage. Although the activation of the prekallikrein-kinin system often remains inapparent, it could affect the arterial pressure or the clotting system in selected subjects, considering that kallikrein is located where many inflammatory precursors join.

Latent and prolonged activation will cause long-term damage because of the continued radical and protease efflux, which silently injures many organs, especially lungs, retina, and central nervous system. Recent reports demonstrated free radicals, proteases, leukotrienes, and other inflammatory products either in plasma or bronchial secretions of stressed neonates (Ogden et al. 1983; Bruce et al. 1984). In samples of bronchial secretions serially collected from critically ill neonates submitted to mechanical ventilation, our group (unpublished data: 1985) detected increasing granulocyte aggregating activity. In the lungs several modes of inflammatory damage can be seen: hypoxemia, PaO_2 fluctuations, and barotrauma due to ventilation alter endothelium and excite both neutrophils and alveolar macrophages. The released radicals peroxidize fatty acids and covalently bound critical sulphydryl groups in membranes, inactivating enzymes and other proteins. Lytic damage, primarily induced by elastase, affects the structural components of connective tissue, such as elastin, collagen, and proteoglycans (Carp and Janoff 1983). When the local balance between arachidonic acid derivatives is disrupted, a prevailing amount of thromboxane and leukotrienes is established, and severe pulmonary hypertension can follow (Fantone and Ward 1983). The recent recognition of these inflammatory events modifies the old concept by which oxidative damage is solely related to hyperoxia, and explains the occurrence of bronchopulmonary dysplasia, retrolental fibroplasia, and other damage in patients whose PaO_2 had been carefully monitored.

Acute and powerful activation of several defense processes contributes to the development of shock, which represents in neonatal age (more than in the subsequent periods) a common terminal point of different severe disorders. The myriad of biological metabolites, proteases, and cellular split products flowing into the blood and tissues deranges the homeostasis which can result in tissue damage involving also the viable organs.

A more restricted, but similarly severe, expression of acute activation is shock lung. This disorder, due to an acute injury of the terminal endothelium-alveolar unit mostly caused by massive intravascular granulocyte aggregation (Rinaldo and Rogers 1982), has been neglected up to now in the neonate, despite the fact that several respiratory disorders often show the typical features of shock lung, such as the pulmonary involvement in group B streptococcus (GBS) infection (Rojas and Stahlman 1984). We first demonstrated shock lung in a premature infant dying from clostridium infection at the 5th day of life. We detected granulocyte aggregating activity (GAA) in plasma and in bronchial secretions concurrently with the onset of respiratory failure and radiological images, which changed from edema to interstitial emphysema, and infiltration spread to both lungs (Laurenti et al. 1985). Shock lung, as an expression of derangement of defense mechanisms, is more common than usually considered and deserves careful attention. It may often excape diagnosis or be misunderstood as atypical RDS.

Since granulocyte aggregation has been recently implicated in various disorders in adults, causing neurological symptoms, sudden blindness, myocardial infarction, and respiratory dysfunction (Hammerschmidt 1983), we planned an extensive study in healthy and stressed newborns in order to investigate the relevance of this phenomenon in the first days of life. In preliminary results (Laurenti et al. 1984), we detected GAA in plasma of many sick neonates (91%), whereas it was totally absent in normal term and preterms. However, no close correlation was observed betwen GAA and clinical manifestations with regard, to severity of symptoms and occurrence of a well-defined syndrome. It is possible that the clinical manifestations excaped, being masked by the underlying disease. So we performed some experiments in order to detect possible phases of total or partial granulocyte deactivation. By challenging sequentially granulocyte suspensions with a primary diluted sample of aggregating plasma and subsequently with the optimal dose, we demonstrated phases of total or subtotal unresponsiveness. These experiments extend those previously made by other authors who documented deactivation after consecutive challenges with fMLP, C5a, and C5a desarginine (Henson et al. 1978; Simkhowitz et al. 1980).

It is possible that some modulating mechanisms resembling those experimentally studied also operate in clinical conditions; therefore, leukostatic disorders are established only when an initial strong impact of aggregating factors on granulocytes occurs. In such instances, a critical element may be the "speed" by which the GAA is generated: a leukoembolization syndrome can develop if GAA increases acutely in plasma, whereas its delayed shedding allows the granulocyte to arrange their circuits of deactivation.

However, apart from a supposed deactivation, neutrophil unresponsiveness can similarly occur for other mechanisms in the stressed neonate. For instance, an imbalance in the generation of free radicals and scavenger factors determines the relevance of damage on the granulocyte membrane due to auto-oxidation (Strauss et al. 1980; Clark 1983).

In summary, different clinical manifestations (i.e., shock lung, respiratory and neurologic disorders) as well as a completely silent clinical picture can follow the activation of the defense mechanisms in the newborn infant. In the latter case, the functional activity of the neutrophils can be suppressed by a series of intricate reactions which involve their routes of activation, deactivation, and auto-oxidation. It is likely that a naturally acquired mechanism acts in the neonate as a protective system against the uncontrolled fluctuation of the defense reactivity.

References

Anderson DC, Pickering LK, Feigin RD (1974) Leukocyte function in normal and infected neonate. J Pediatr 85:420–425

Anderson DC, Hughes BJ, Smith CW (1981) Abnormal motility of neonatal polymorphonuclear leukocytes. J Clin Invest 68:863–874

Ambruso DR, Bentwool B, Henson PM, Johnston RB (1984) Oxidative metabolism of cord blood neutrophils: relationship to content and degranulation of cytoplasmic granules. Pediatr Res 18:1148–1153

Bruce MC, Martin RJ, Boat TF (1984) Concentrations of alpha 1-proteinase inhibitor and alpha2-macroglobulin in serum and lung secretions of intubated infants. Pediatr Res 18:35–40

Carmona R, Catalano R, Trunkey DD (1984) Septic Shock In: Shire GT (ed), Shock and related problems, p. 156–177, Churchill Livingstone, London

Carp H, Janoff A (1983) Modulation of inflammatory cell protease-tissue antiprotease interaction at sites of inflammation by leukocyte-derived oxidants. In Weissmann G (ed), Advances in Inflammatory Research, vol. 5, p. 173–202, Raven Press N. Y.

Christensen RD, Rothstein G (1980) Exhaustion of mature marrow neutrophils in neonates with sepsis. J. Pediatr 96:316–318

Clark RA (1983) Extracellular effects of the myeloperoxidase-hydrogen peroxide-halide system In Weissmann G (ed), Advances in Inflammatory Research, vol 5, p. 107–146, Raven Press N. Y.

Cocchi P, Mori S, Becattini A (1971) NBT reduction by neutrophils of newborn infants in vitro phagocytosis test. Acta Pediatr Scand 60:475–478

Dahiden C, Galanos C, Fehr J (1983) Granulocyte activation by endotoxin. J Immunol 130:857–863

Davies PA, Gothefors LA (1984) Bacterial infections in the fetus and newborn infant (Major Problem in Clinical Pediatrics), Saunders, London

Demopoulos HB, Flamm ES, Pietronigro DD, Seligman ML (1980) The free radical pathology and the microcirculation in the major central nervous system disorders. Acta Physiol Scand, Suppl 492:91–120

Fantone JC, Ward PA (1983) Mechanisms of neutrophil-dependent lung injury In: Ward PA (ed), Immunology of Inflammation (Series Handbook of Inflammation, vol 4), p. 89–119, Elsevier Science Publ, Amsterdam

Franck L and Groseclose EE (1984) Preparation for birth into an O_2 rich environment: the antioxidant enzymes in the developing rabbit lung. Pediatr Res 18:240–244

Frazier JP, Cleany TG, Pickering LK, Hohl S, Ross PJ (1982) Leukocyte function in healthy neonates following vaginal and cesarean section deliveries. J Pediatr 101:269–272

Goldstein IM (1984) Neutrophil degranulation In: Snyderman R (ed), Regulation of leukocyte function (Contemporary Topics in Immunobiology) vol 14, p. 189–220, Plenum Publ Corp, N. Y.

Hammerschmidt DE (1983) Activation of the complement system and of granulocytes in lung injury: the adult respiratory distress syndrome In: Weissman G (ed), Advances in Inflammatory Research, vol 5, p. 147–174, Raven Press, N.Y.

Hammerschmidt DE, Ogburn PL, Williams JE (1984) Amniotic fluid activates complement. J Lab Clin Med 104:901–907

Henson PM, Zanolari B, Schwartzman NA (1978) Intracellular control of human neutrophil secretion I- C5a induced stimulus-specific desensibilization and the effect of cytochalasin B. J Immunol 121:851–858

Laurenti F, Ferro R, Marzetti G, Rossini M, Bucci G (1980) Neutrophil chemotaxis in preterm infants with infections. J Pediatr 96:468–470

Laurenti F, Fiorucci P, Pacifico L, Turrini S, Messina E, Rossini M, Moretti C, Bucci G (1984) Granulocyte aggregation in neonate. Pediatr Research 18, abs.

McCord JM (1985) Oxygen-derived free radicals in post-ischemic tissue injury. N Eng J Med 312:159–163

McPhail LC, Snyderman R (1984) Mechanisms of regulating the respiratory burst in leukocytes In: Snyderman R (ed), Regulation of leukocyte function (Contemporary Topics in Immunobiology) vol 14, p. 247–269, Plenum Press N.Y.

Miller ME (1978) Host Defenses in the Human Neonate (Monograph in Neonatology), Grune & Stratton, N.Y.

Ogden BE, Rushton A, Johnson JD, Murphy SA (1983) Bronchoalveolar lavage factor in RDS and BPD. Pediatr Res 17, abs.

Rinaldo JE, Rogers RM (1982) ARDS: changing concepts of lung injury and repair. N Eng J Med 306:900–909

Rojas J, Stahlman M (1984) The effect of group B Streptococcus and other organisms on the pulmonary vasculature. Clin Perinat 11:591–599

Sangstad OD, Harvie A, Langslet A (1982) Activation of the kallikrein/kinin system in premature infants with RDS. Acta Pediatr Scand 71:956–958

Salem HH, Walters WA, Perkin JL, Handley CI, Firkin BG (1982) Aggregation of human platelets by amniotic fluid. Br J Obst Gynec 89:733–737

Scheifele DIV, Melton P, Whitchelo V (1981) Evaluation of the Limulus test for endotoxemia in neonates with suspected sepsis. J Pediatr 98:899–903

Shigeoka AO, Santos JI, Hill HL (1979) Functional analysis of neutrophil granulocytes from healthy, infected and stressed neonates. J Pediatr 95:454–460

Shigeoka AO, Clarette RP, Wyman ML, Hill HR (1981) Defective oxidative metabolic responses of neutrophils from stressed neonates. J Pediatr 98:392–398

Simkowitz L, Atkinson JP, Spilgerg I (1980) Stimulus-aspecific deactivation of chemotactic factor-induced cAMP response and superoxide generation by human neutrophlis. J Clin Invest 66:736–741

Sklar LA, Jesaitis AJ, Painter RG (1984) The neutrophil N-Formyl-Peptide Receptor: Dynamic of ligand-receptor interactions and their relationship to cellular responses – In: Snyderman G (ed), Regulation of Leukocyte Function (Contemporary Topics in Immunobiology) vol 14, p. 29–74, Plenum Press N.Y.

Stenmark KR, James SL, Voelkel NF (1983) Leukotriene C4 and D4 in neonates with hypoxemia and pulmonary hypertension. N Eng J Med 308:77–80

Strauss RG, Hart MJ (1981) Spontaneous and drug induced ConA capping of neutrophils from human infants and their mothers. Pediatr Res 15:1314–1318

Strauss RG, Snyder EL (1983) Activation and activity of the superoxide-generating system of neutrophils from infants. Pediatr Res 17:662–664

Strauss RG, Snyder EL, Wallace PD, Rosenberg TG (1980) Oxygen detoxifying enzymes in neutrophils of infants and their mothers. J Lab Clin Med 95:897–903

Yurt RW (1984) Intermediary metabolism including mediator activation. In: Shires GT (ed), Shock and Related Problems, p. 111–126, Churchill Livingstone, London

Wright WC Jr, Ank BJ, Herbert J, Stiehm ER (1975) Decreased bactericidal activity of leukocytes of stressed newborn infants. Pediatrics 56:579–584

T-Cell Development and Function: Relationship to Immunodeficiencies

F. Aiuti, R. Paganelli, B. Ensoli, M. Crescenzi, M. Carbonari, and M. Fiorilli

Department of Allergy and Clinical Immunology, University of Rome *La Sapienza*, Rome, Italy

1 T-Cell Development and Function

Recent advances in knowledge about the phenotypical and functional aspects of T-cell ontogeny, as well as about the molecular events involved in this process, promise to be of considerable help in the understanding of several human immunodeficiency disorders. The earlier steps of intrathymic differentiation have recently been shown to involve the expression of surface molecules, such as the receptors for interleukin-2 (IL-2) and the T11 determinant, which might be of paramount importance as receptors for inductive signals (von Boehmer et al. 1985; Meuer et al. 1984). These signals would induce, within the thymic environment, further maturation of early thymocytes. The stepwise differentiation of these cells is marked by the acquisition of discrete surface markers and by rearrangements of the genes encoding for T-cell receptor structures. The transition from early (stage I) to common (stage II) thymocytes is marked by the appearance of the T6 determinant and by the rearrangement of the T-cell receptor (TCR) beta-chain genes (Royer et al. 1984). The further transition to stage III thymocytes is characterized by the disappearance of T6, the rearrangement of TCR alpha-chain genes, and the surface expression of functional TCRs bound to the T3 molecule (T3Ti complex) (Royer et al. 1984; Collins et al. 1985). Virtually nothing is known about the functional role of the T-cell receptor-like structure encoded by the gamma gene (Saito et al. 1984). The possibility that this molecule might be involved in self-recognition of major histocompatibility complex (MHC) during the earliest stages of intrathymic differentiation is interesting (Raulet et al. 1985).

Most patients with T-cell defects (e.g., DiGeorge's syndrome or thymic hypoplasia) have normal natural killer (NK) cell activity. These findings, together with the observation of normal NK activity in *nude* or neonatally thymectomized mice, led to the conclusion that the functional development of NK cells does not depend strictly on the thymic environment. The recent division of cells with NK activity into distinct subsets led to a revision of this assumption. In fact, NK cells appear to be broadly divisible into two subpopulations: one bears certain T-cell-specific surface determinants (the 90-kd Ti heterodimer and, in humans, the T3 determinant), and is characterized by the functional rearrangement of TCR alpha- and beta-chain genes; the other subpopulation lacks surface T3Ti and does not express full-length TCR beta transcripts (Yanagy et al. 1985; Ritz et al.

Immunology of the Neonate
Burgio, Hanson, Ugazio (Eds.)

1985). Interestingly, however, the T3Ti-negative NK cells express a truncated TCR beta-chain messenger ribonucleic acid (mRNA), suggesting a relationship of these cells with early T cell precursors (Ritz et al. 1985). It might be of interest to examine the rearrangement and expression of TCR genes in NK cells from congenitally athymic individuals.

An interesting form of immunodeficiency is that resulting from defective expression of the MHC products - class I, class II, or both classes of molecules (Touraine et al. 1978; Griscelli, personal communication, 1985). Patients have been shown by Southern blotting to possess normal genes encoding for class I or II molecules, but to fail to express the corresponding mRNAs even after in vitro induction with gamma interferon (IFN-γ). (C. Griscelli, personal communication, 1985). Thus, the defect in these patients is probably due to abnormal regulation of MHC gene expression. Clinically, these patients are characterized by immunodeficiency with lack of in vivo and in vitro immune response to exogenous antigens. It is likely that this deficiency depends on the failure of antigen presentation in the context of class II MHC molecules.

2 IL-2/IL2-R Regulation in Primary Immunodeficiencies

Several aspects of defective T-cell function in primary immunodeficiency diseases have been revised after the discovery of soluble mediators (interleukins) that regulate the proliferation and differentiation of T and B cells. We (Aiuti et al. 1984) investigated the balance of interleukin-2 (IL-2) production and IL-2 receptor (IL2-R) expression in several T-cell-deficient patients in an attempt to determine at what level the alteration(s) leading to hyporesponsiveness to mitogenic stimuli occur. In fact, the exogenous addition of recombinant IL-2 (rIL-2; kindly provided by Biogen, Geneva, Switzerland) was not able to restore proliferation after stimulation with optimal concentrations of phytohemagglutinin (PHA-P, Wellcome; 1 μg/ml) in the majority of patients with severe T-cell defects such as severe combined immune deficiency (SCID), ataxia-telangiectasia (AT), hyper-immunoglobulin E (IgE) syndrome, or Wiskott-Aldrich (WA) syndrome (Table 1). The expression of the IL2-R after 72-h culture in the presence of PHA was evaluated in the same patients by reactivity of the cultured cells with the monoclonal antibody anti-Tac (kindly provided by T. A. Waldmann), which recognizes an epitope of the IL2-R (Leonard et al. 1982). IL2-R expression was depressed in all these patients (Table 1); furthermore, unlike with normal cells, IL2-R expression in patients' cells was not up-regulated by the presence of exogenous rIL-2 (data not shown). Thus, we were able to conclude that deficient mitogen responsiveness in these patients depended substantially on altered IL2-R expression. Functionally mature T-cells were, however, present in less severe cases, such as a minority of common variable immunodeficiency (CVI) cases, patients with primary T-cell defects, and other patients. Here, a significant increase of proliferation was observed upon addition of rIL-2 or crude IL-2 preparations; furthermore, a significant reduction of cells expressing IL2-R, in comparison with normal controls, was seen in the majority of these patients. Thus, in

Table 1. Proliferation and IL2-R expression in response to PHA stimulation with and without rIL-2 by T-cells from immunodeficient patients

Diagnosis	Proliferative response (cpm)		% IL2-R cells
	PHA	PHA+rIL-2	PHA
SCID	5 552	6 830	2
SCID	942	2 548	9
SCID	5 869	3 574	20
Hyper-IgE	18 531	12 590	15
Hyper-IgE	1 304	5 022	35
WA	8 890	15 224	19
WA	779	670	2
AT	5 696	6 204	22
AT	664	814	17
Controls	> 18 000	> 20 000	61 ± 15

these cases poor proliferative responses appeared to depend on a combination of defective IL-2 production and IL2-R expression.

3 IFN-γ Production in Primary Immunodeficiencies

In vitro production of IFN-γ, a T-cell-derived lymphokine with potent immunomodulatory properties (Baron et al. 1982), was found to be grossly defective in AT patients (Paganelli et al. 1984) and in other severe T-cell defects (SCID, WA syndrome, Hyper-IgE syndrome) which is in contrast to the lack of major defects of IFN-γ production in most patients with CVI. In three selected cases with no production of IFN-γ after induction with PHA, staphylococcal enterotoxin B, or galactose oxidase, we found that the defect was not secondary to absence of IL-2, since the addition of rIL-2 did not restore the secretion of IFN-γ (Table 2). Also, the addition of a calcium ionophore (A23187) to the cultures, as IFN inducer, failed to restore IFN-γ production and ruled out a defect of calcium influx into the cells after mitogen stimulation. In fact, no other membrane defect was found

Table 2. IFN-γ production (units/ml) by T-cells from immunodeficient patients

	Stimulus						
Diagnosis	NDV	None	rIL-2	SEB	SEB+rIL-2	GO	GO+rIL-2
AT	2 000	< 3	< 3	< 3	10	< 3	10
SCID	10 000	< 3	< 3	3	30	< 3	< 3
Hyper-IgE	3 000	< 3	< 3	< 3	3	< 3	< 3
Control	10 000	< 3	< 3	300	300	100	200

NDV, Newcastle disease virus (IFN-α); *SEB*, staphylococcus enterotoxin B; *GO*, galactose oxidase

(data not shown). This seems to exclude a defective macrophage-T-cell interaction and/or alteration of the cellular membrane as the basis for the deficiency of IFN-γ in these patients. In addition, all patients showed normal production of IFN-α when stimulated with Newcastle disease virus.

4 Genetic Basis for Immunodeficiency in Ataxia-Telangiectasia

Ataxia-telangiectasia is a rare autosomal recessive disorder characterized by cerebellar ataxia, oculocutaneous telangiectasia, variable immunodeficiency, high incidence of cancer, increased chromosomal breakage, and hypersensitivity to DNA-damaging agents (Bridges and Harnden 1982). The genetic lesion or lesions leading to defective DNA repair and to the pleiotropic clinical manifestations of AT are still obscure.

Recently, heterogeneity of AT has become apparent on the basis of genetic complementation studies (Jaspers and Bootsma 1982) and of clinical and immunological data (Fiorilli et al. 1983). Further heterogeneity within this syndrome is suggested by the recent observation (Fiorilli et al. 1985a) of a distinct subset of patients characterized by low-level radiosensitivity and minimal immunological impairment. Cells from these patients differ from those of patients with "classical" AT in their lower frequency of spontaneous or bleomycin-induced chromosomal breakage and their ability to normally inhibit DNA synthesis following ionizing irradiation. The possibility of a relationship of DNA-repair deficiency with the clinical manifestations of AT arises from the observation that patients with "low-radiosensitive" variant AT appear to be only mildly immunologically compromised (Fiorilli et al. 1985a). This suggests a direct relationship between the DNA-repair defect and the damage to the immune system in this disease.

T and B lymphocytes are remarkably similar in the way they generate the clonal diversity of their antigen receptor structures. In fact, both types of cells use somatic rearrangement and mutation of limited germ-line genetic information encoded by variable (V), diversity (D), joining (J), and constant (C) regions (Patten et al. 1984; Tonegawa 1983). These rearrangements require the cutting and ligation of specific sequences flanking V, D, J, and C genes. Clearly, these recombinational events must be regulated by a set of specific enzymes; while V-J region-specific nucleases have been identified (Desiderio and Baltimore 1984), virtually nothing is known about the enzymes involved in the rejoining of immunoglobulin or T-cell receptor genes during physiological rearrangements. The spontaneous chromosomal rearrangements in AT lymphocytes occur nonrandomly at specific regions, namely 14q32, 14q11–12, 7q35, and 7p14 (Bridges and Harnden 1982). Interestingly enough, the recent mapping of the T-cell receptor alpha-chain genes at 14q11 (Croce et al. 1985), and of the beta- and gamma-chain genes at 7q35 and 7p12 respectively (Murre et al. 1985), has confirmed the previous suggestion (Fiorilli et al. 1985b) that these nonrandom breaks occur at regions encoding for genes which undergo rearrangements in lymphoid cells. As the increased frequency of chromosomal breaks in AT appears to be due to a defect of rejoining (Collins et al. 1985), it can reasonably be argued that the same recombinase (which is defective in AT) involved in the repair of radiogenic damage also mediates the physiological rearrangements of lymphocyte receptor genes. Defective ligation might, therefore, lead to an increased frequency of ineffective rearrangements of T- and B-cell receptor genes, resulting in the death of the majority of lymphoid cells.

References

Aiuti F, Paganelli R, Le Moli S, Seminara R, Cherchi M (1984) Defects of interleukins in immunodeficiencies. In: Griscelli C, Vossen J (eds) Progress in immunodeficiency research and therapy. Elsevier, Amsterdam, pp 261–266

Baron S, Dianzani F, Stanton GJ (eds) (1982) The interferon system. Tex Rep Biol Med 41:1–715

Bridges BA, Harnden DG (eds) (1982) Ataxia-telangiectasia. Wiley, Chichester

Collins MKL, Tanigawa G, Kissonerghis AM, Ritter M, Price KM, Tonegawa S, Owen NJ (1985) Regulation of T cell receptor gene expression in human T cell development. Proc Natl Acad Sci USA 82:4503–4507

Croce CM, Isobe M, Palumbo A, Puck J, Ming J, Tweardy D, Erikson J, Davis MM, Rovera G (1985) Gene for alpha-chain of human T cell receptor: location on chromosome 14 region involved in T cell neoplasms. Science 227:1044–1047

Desiderio S, Baltimore D (1984) Double-stranded cleavage by cell extracts near recombination signal sequences of immunoglobulin genes. Nature 308:860–862

Fiorilli M, Businco L, Pandolfi F, Paganelli R, Russo G, Aiuti F (1983) Heterogeneity of immunological abnormalities in ataxia-telangiectasia. J Clin Immunol 3:135–141

Fiorilli M, Antonelli A, Russo G, Carbonari M, Crescenzi M, Petrinelli P (1985a) Variant of ataxia-telangiectasia with low-level radiosensitivity. Hum Genet 70:274–277

Fiorilli M, Carbonari M, Crescenzi M, Russo G, Aiuti F (1985b) T cell receptor genes and ataxia-telangiectasia. Nature 313:186

Jaspers NGJ, Bootsma D (1982) Genetic heterogeneity in ataxia-telangiectasia studied by cell fusion. Proc Natl Acad Sci USA 79:2641–2644

Leonard WJ, Depper JM, Uchiyama T, Smith KA, Waldmann TA, Greene WC (1982) A monoclonal antibody that appears to recognize the receptor for human T cell growth factor; partial characterization of the receptor. Nature 300:267–269

Meuer SC, Hussey RE, Fabbi M, Fox D, Acuto O, Fitzgerald KA, Hodgdon JC, Protentis JP, Schlossman SF, Reinherz EL (1984) An alternative pathway of T cell activation: a functional role for the 50 kd T 11 sheep erythrocyte receptor protein. Cell 36:897-906

Murre C, Waldmann RA, Morton CC, Bongiovanni KF, Waldmann TA, Shows TB, Seidman JG (1985) Human gamma-chain genes are rearranged in leukemic T cells and map to short arm of chromosome 7. Nature 316:549–552

Paganelli R, Capobianchi MR, Matricardi PM, Cioé L, Seminara R, Dianzani F, Aiuti F (1984) Defective interferon-gamma production in ataxia-telangiectasia. Clin Immunol Immunopathol 32:387–391

Patten P, Yokota T, Rothbard J, Chein Y, Arai K, Davis MM (1984) Structure, divergence and expression of T cell receptor beta-chain variable regions. Nature 312:40–46

Raulet DH, Garman RD, Saito H, Tonegawa S (1985) Developmental regulation of T cell receptor gene expression. Nature 314:103–107

Ritz J, Campen TJ, Schmidt RE, Royer HD, Hercend T, Hussey RE, Reinherz EL (1985) Analysis of T cell receptor gene rearrangement and expression in human natural killer clones. Science 228:1540–1544

Royer HD, Acuto O, Fabbi M, Tizard R, Ramachandran K, Smart JE, Reinherz EL (1984) Genes encoding the Ti beta subunit of the antigen/MHC receptor undergo rearrangement during intrathymic ontogeny prior to surface T3-Ti expression. Cell 39:261–266

Saito H, Kranz DM, Takagaki Y, Hayday AC, Eisen HN, Tonegawa S (1984) Complete primary structure of a heterodimeric T cell receptor deduced from cDNA sequences. Nature 309:757–762

Tonegawa S (1983) Somatic generation of antibody diversity. Nature 302:575–581

Touraine JL, Betuel H, Suillet G, Jeune MJ (1978) Combined immunodeficiency associated with absence of cell surface HLA A and B antigens. J Pediatr 93:47–50
von Bohemer H, Crisanti A, Kisielow P, Haas W (1985) Absence of growth by most receptor-expressing fetal thymocytes in the presence of interleukin-2. Nature 314:539–540
Yanagy Y, Caccia N, Kronenberg M, Chin B, Roder I, Rohel B, Kiyohara T, Lauzon R, Toyonaga B, Rosenthal K, Dennert G, Acha-Orbea H, Hengartner H, Hood L, Mak TW (1985) Gene rearrangement in cells with natural killer activity and expression of the beta-chain of the T cell antigen receptor. Nature 314:631–633

Activation and Functional Capacity of Neonatal T Cells: Analysis of Interleukin-2 and Interleukin-2 Receptor Responsiveness

DAVID L. NELSON

Immunophysiology Section, Metabolism Branch, National Cancer Institute, National Institutes of Health Building 10, Room 4N112, Bethesda, MD 20892, USA

1 Introduction

The thymic epithelium is first recognized in the human fetus near the 6th week of fetal gestation. This epithelial anlage becomes populated with lymphocytes which express T-lineage cell surface molecules during the 9th to 10th weeks, and these cells become responsive to activating stimuli during the 10th to 12th weeks of gestation. Such T lymphocytes are found in the spleen 3–4 weeks after their appearance in the thymus and subsequently enter the fetal circulation (reviewed by Stiehm et al. 1979; Andersson et al. 1981; Hayward 1981; Toivanen et al. 1981). In spite of the early appearance of such T cells during gestation, our most extensive knowledge of the immunological capabilities of early human T cells is derived from studies of human neonatal T cells derived from cord blood at the time of delivery. These cord blood T cells have been stimulated or activated in a number of ways and their functional capabilities assessed. Cord blood T cells have been shown to proliferate in response to polyclonal T-cell activators as well as certain foreign antigens; to provide help for immunoglobulin production by B cells in response to polyclonal B-cell activators and antibody responses to specific antigens; to exert suppressor influences on B cells and T cells through cellular mechanisms and the release of soluble suppressor molecules; to mediate cytotoxic responses to specific antigens and when activated with polyclonal T-cell activators; and to release a variety of lymphokines. One of these lymphokines is termed interleukin-2 (IL-2) and was previously termed T-cell growth factor owing to its ability to support the growth of activated T cells following interaction with specific cellular receptors for IL-2 (IL-2R). Currently, studies in a variety of areas are delineating the critical roles played by this lymphokine and its receptor in the development, maturation, and immunoregulation of the human immune response (Cantrell and Smith 1984). IL-2, which supports the proliferation of activated T cells and B cells (Waldmann et al. 1984), is secreted in normal amounts by phytohemagglutinin (PHA)-activated neonatal T cells (Hayward and Kurnick 1981; Wakasugi et al. 1985), and neonatal T cells also proliferate to lectin-depleted IL-2 preparations (Hayward and Kurnick 1981). Using two hybridoma-derived monoclonal antibodies directed against the human IL-2R, anti-Tac (Uchiyama et al. 1981) and 7G7B6 (Rubin et al. 1985a), our laboratory has developed a "sandwich" enzyme-linked immunosorbent assay (ELISA) to quantitatively measure human IL-2R in solution (Rubin et al. 1985b). Using this ELISA

Immunology of the Neonate
Burgio, Hanson, Ugazio (Eds.)

for soluble IL-2R, we have shown with adult volunteers that: (a) IL-2R can be quantitated rapidly in soluble form in the cell-free supernatants of activated adult T cells and B cells as well as detergent-solubilized extracts of these same cells; (b) supernatant and cell-associated IL-2R appearance is rapid (<24 h), requires protein synthesis but not cellular division, and results from both nonspecific and specific (antigenic) stimulation of cells; (c) the released form of the IL-2R possesses a molecular weight 10 000 daltons smaller than the cell-associated receptor; and (d) the released IL-2R is capable of specifically binding to its ligand, IL-2, and is present in the serum and plasma of normal individuals (Rubin et al. 1985b). Such a soluble form of the IL-2R might serve as a sensitive marker of immune activation in vitro and in vivo and exert potentially important immunoregulatory functions. An analysis of neonatal T cells for the capacity of generating supernatant and cell-associated IL-2R was therefore undertaken as well as an analysis of cord blood plasma for levels of soluble IL-2R.

2 Materials and Methods

2.1 Preparation of Cord Blood and Adult Peripheral Blood

Mononuclear Cells
Cord blood mononuclear cells (CBMC) from full-term fetuses of uncomplicated pregnancies and adult peripheral blood mononuclear cells (PBMC) were prepared as described (Yarchoan and Nelson 1983). Cord blood plasma and adult plasma were prepared by centrifugation of heparinized whole blood.

2.2 Antibodies

The murine monoclonal anti-IL-2R antibody termed anti-Tac (Uchiyama et al. 1981) was provided by Dr. T. A. Waldmann, National Institutes of Health. The murine monoclonal anti-IL-2R, 7G7B6, which binds to the human IL-2R at an epitope distinct from anti-Tac and IL-2, was produced and modified with fluorescein isothiocyanate (FITC) as described elsewhere (Rubin et al. 1985a). The affinity-purified, alkaline phosphatase-conjugated rabbit anti-FITC was produced as described (Yarchoan et al. 1984).

2.3 Cell Cultures and Stimulants

Cultures of CBMC and adult PBMC, containing 1×10^6 mononuclear cells in 2 ml RPMI 1640 supplemented with 10% fetal calf serum (FCS), 100 units/ml penicillin, and 100 μg/ml streptomycin, were established in the wells of 24-well plates. Triplicate cultures were incubated for various times in a humidified atmosphere of 5% CO_2 in air at 37° C. Cultures were either unstimulated (medium alone) or they contained PHA (GIBCO, Grand Island, NY) at a final dilution of

1/200, the murine monoclonal antibody OKT3 (Ortho, Raritan, NJ) at a final concentration of 10 ng/ml, pokeweed mitogen (PWM) (GIBCO) at a final concentration of 1/100, or delectinated natural IL-2 (Cellular Products, Buffalo, NY) at a concentration of 10% (v/v). Cultures were harvested by centrifugation at 550 *g* for 10 min and the supernatants removed. The cell pellet was washed once in Dulbecco's phosphate-buffered saline (PBS) containing 25 m*M* Tris, pH 7.4 (TBS). The cell pellet was solubilized by incubation for 15 min in 0.5 ml of TBS containing 1% Triton X-100 (Research Products International, Elk Grove, Ind), followed by centrifugation in a microfuge (Eppendorf) and recovery of the supernatant. Culture supernatants and detergent-solubilized extracts were stored at –20° C until analyzed for IL-2R.

2.4 ELISA Assays

The ELISA for soluble IL-2R was performed as described in detail elsewhere (Rubin et al. 1985b). The ELISA is schematically shown in Fig. 1. The ladder of reagents from the solid phase up consisted of anti-Tac, soluble IL-2R, FITC-conjugated 7G7B6, alkaline phosphatase-conjugated rabbit anti-FITC, and the substrate p-nitrophenyl phosphate. The conversion of substrate to product was measured spectrophotometrically at 405 n*M*. The absorbance of the experimental wells was compared against standard curve (Fig. 1) generated by adding varying amounts of IL-2R as previously described in detail (Yarchoan et al. 1981). The IL-2R standard was the cell-free supernatant of an in vitro-passaged T-cell line which was assigned a level of 1000 IL-2R units/ml. As can be seen, for any given absorbance value, a numerical value for the quantity of IL-2R can be derived. The ELISA for IgM was performed as described elsewhere (Yarchoan et al. 1982).

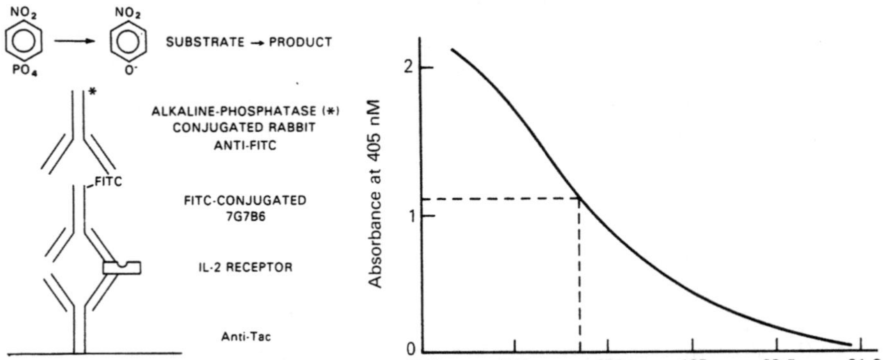

Fig. 1. Schematic of the ELISA employed for the measurement of soluble IL-2R (see *Materials and Methods)*

2.5 Assessment of Proliferative Responses

Proliferative responses of cultured cells were estimated by pulsing 200-μl aliquots of cell suspension obtained at the time of harvesting with 1 μCi [^{3}H]thymidine (New England Nuclear, Boston, Mass) and harvesting as described (Nelson et al. 1983).

2.6 Sera from Patients with Cancer

Sera were obtained retrospectively from patients with various forms of neoplasia from the serum bank of the Metabolism Branch of the National Cancer Institute, NIH. Control sera were similarly obtained.

2.7 Calculations

Geometric means and SEM were calculated for the responses of groups of donors, and those values less than 62.5 units of IL-2R were assigned a value of 62.5 for the purposes of calculations.

3 Results

3.1 Cellular and Supernatant IL-2R Generation by CBMC and PBMC with Various Activators

In previous studies of IL-2R generation, we found that activated adult PBMC generated the most cell-associated IL-2R between days 3 and 5 of culture, while soluble supernatant IL-2R levels peaked on about day 7 of culture (Rubin et al. 1985b). Therefore, we first examined the cellular IL-2R production of activated CBMC and adult PBMC on day 5 of culture, and supernatant IL-2R production on day 7 of culture. As can be seen in the left panel of Fig. 2, all CBMC and adult PBMC expressed cellular IL-2R on day 5 when stimulated with PHA or OKT3, and an occasional CBMC and PBMC sample made low levels of measurable receptor when cultured with medium alone.

There was no detectable IL-2R expression (>62.5 units/ml) in any of the PBMC or CBMC samples when analyzed for IL-2R when freshly isolated and prior to culture in vitro. For samples stimulated with IL-2 alone, the cell-associated IL-2 levels on day 5 were: CMBC 465 $\pm$ 1.5 (mean $\pm$ 1 SEM) with a range of 108–1801; adult PBMC 593 $\pm$ 1.6 with a range of 179–2498. There were no statistically significant differences between CBMC and adult PBMC for any stimulus tested. Supernatants harvested at day 5 also failed to show any significant differences between CBMC and adult PBMC for any activator examined (data not shown). When supernatants from these same donors were analyzed for soluble IL-2R on day 7 (Fig. 2, right panel), it can be seen that each donor made soluble

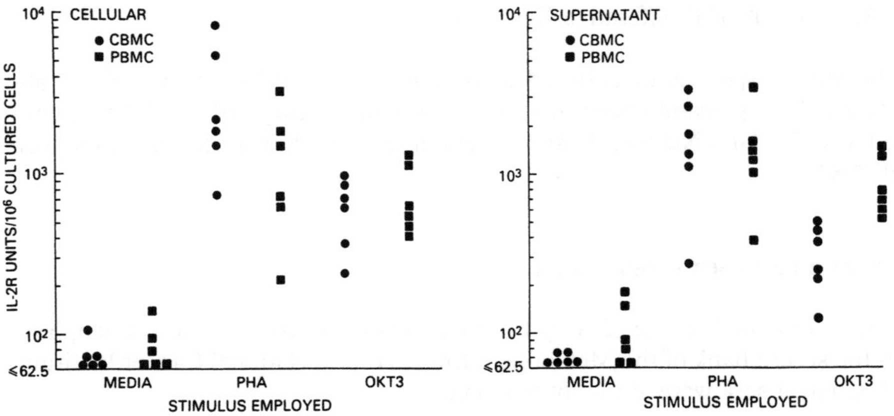

Fig. 2. Expression of cell-associated *(left panel)* and supernatant *(right panel)* IL-2R by CBMC and adult PBMC following in vitro activation. Cultures of adult PBMC and neonatal CBMC were activated with tissue culture media alone, PHA, or the murine monoclonal antibody OKT3. Cellular IL-2R *(left panel)* was assayed on day 5 and supernatant IL-2R was examined on day 7. Results are expressed in units of IL-2R per million cultured cells

IL-2R with PHA and OKT3 as stimuli, and, again, some donors made measurable supernatant IL-2R when cultured with medium alone.

The mean supernatant levels of soluble IL-2R in cultures stimulated with delectinated natural IL-2 were: CBMC 734 ± 1.1 (range 505–1073); adult PBMC 914 ± 1.4 (range 389–2016). Estimates of the levels of supernatant IL-2R generated by CBMC and adult PBMC in response to this preparation of IL-2 were complicated by the fact that the undiluted delectinated preparation of natural IL-2 contained roughly 2000 units/ml soluble IL-2R and, thus, when used at a 10% concentration, would contribute 200 units/ml of IL-2R to these supernatants. Therefore, only that amount of receptor over 200 units/ml would be contributed by the cells in culture. The only significant difference noted between CBMC and PBMC was for supernatant IL-2R production induced by OKT3 ($p < 0{,}01$).

3.2 Time Course of Cellular and Supernatant IL-2R Production Following Cellular Activation

While these studies of IL-2R in supernatants on day 7 and cell-associated IL-2R on day 5 suggested that neonatal mononuclear cells were competent for the production of IL-2R, we undertook experiments to see whether the time course of cellular and supernatant IL-2R production might be different for CBMC versus adult PBMC. An experiment involving PHA-activated cells is shown in Fig. 3. As can be seen, the expression of cellular IL-2R at various times following activation was similar in CBMC and adult PBMC, and the generation of supernatant IL-2R in these same cultures (Fig. 3, right panel) was also comparable. Similarly, no differences in the time course of cellular and supernatant IL-2R

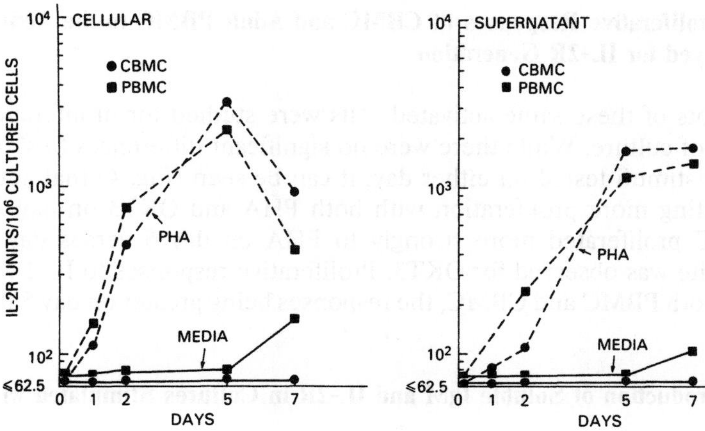

Fig. 3. Time course of cell-associated and supernatant, soluble IL-2R following PHA activation of CBMC and adult PBMC. For details see legends of Figs. 1 and 2

generation were observed in cultures of CBMC stimulated with IL-2, but significantly less supernatant (but not cellular) IL-2R production was again observed in OKT3-stimulated CBMC compared to adult PBMC on days 1, 2, 5, and 7 (data not shown).

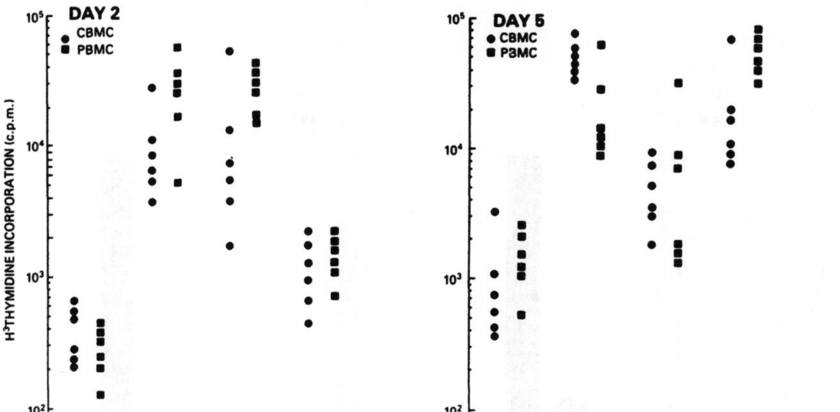

Fig. 4. Proliferative responses of CBMC and adult PBMC to media alone, PHA, OKT3, and delectinated natural IL-2. Proliferative responses were estimated by [^{3}H]thymidine incorporation on day 2 *(left panel)* and day 5 *(right panel)*

3.3 Proliferative Responses of CBMC and Adult PBMC to the Activating Stimuli Employed for IL-2R Generation

Aliquots of these same activated cells were studied for proliferation on days 2 and 5 of culture. While there were no significant differences ($p < 0.01$) with any of the stimuli tested on either day, it can be seen (Fig. 4) that adult cells were exhibiting more proliferation with both PHA and OKT3 on day 2 than day 5. CBMC proliferated more strongly to PHA on day 5 versus day 3, while the opposite was observed for OKT3. Proliferative responses to IL-2 were observed with both PBMC and CBMC, the responses being greater on day 5 than day 2.

3.4 Production of Soluble IgM and IL-2R in Cultures Stimulated with PWM

Numerous studies have documented that CBMC fail to produce IgM when stimulated with PWM, in part because of the presence of T-suppressor cells (reviewed by Tosato et al. 1980). To examine whether or not a similar phenomenon might be observed for IL-2R, we stimulated CBMC and adult PBMC with PWM and analyzed the supernatants for IgM and IL-2R by ELISA. As can be seen in Fig. 5, PBMC and CBMC both made comparable levels of IL-2R when stimulated with PWM, but only the adult PBMC made supernatant IgM.

3.5 Levels of IL-2R in Cord Blood Plasma, Adult Plasma, and the Serum from Normal Individuals and Patients with Cancer

Since normal adult individuals posses measurable levels of serum IL-2R (Rubin et al. 1985b), we determined the levels of IL-2R in plasma derived from whole

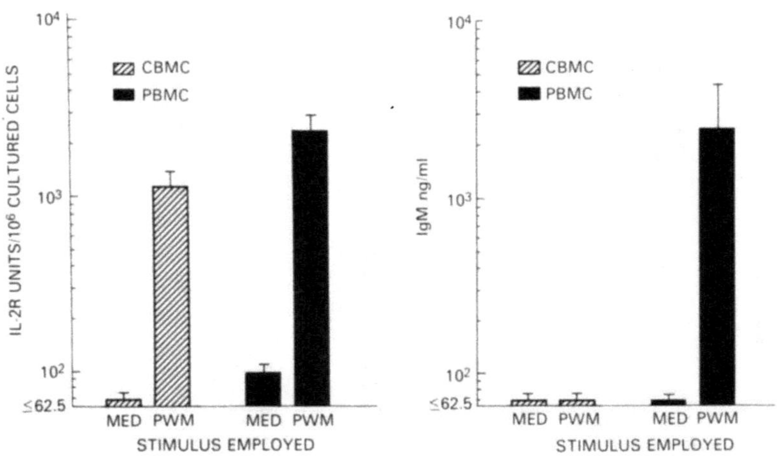

Fig. 5. Production of supernatant IL-2R *(left panel)* and IgM *(right panel)* following the in vitro activation of CBMC and adult PBMC with PWM. Results are expressed as nanograms/milliliter for IgM and as IL-2R units per million cultured cells for soluble IL-2R

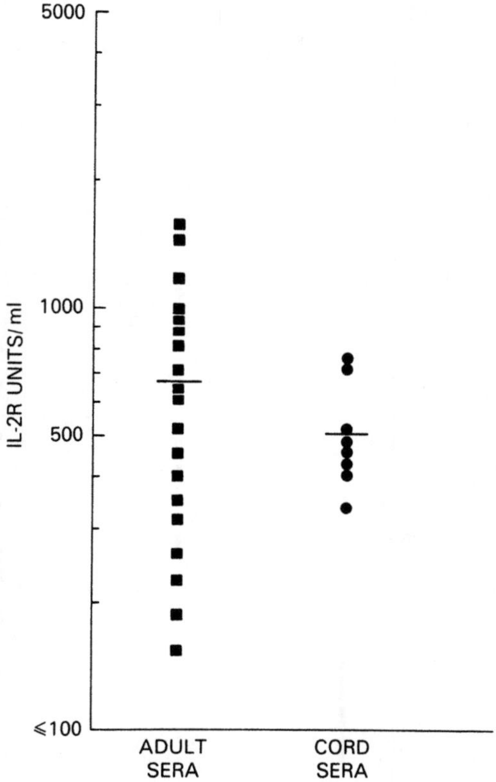

Fig. 6. Plasma levels of IL-2R from cord blood and normal adult controls. Results represent the levels of IL-2R in units per milliliter of plasma. The *horizontal bar* represents the mean value for each group of plasmas

cord blood and compared them with those in normal adult plasma. As can be seen in Fig. 6, each of eight cord blood plasmas contained detectable levels of IL-2R, and the levels were not significantly different from that of adult plasma. These in vitro and in vivo results suggested that neonatal lymphocytes, which might represent cells at relatively early states of maturation and/or differentiation, were fully capable of releasing soluble IL-2R. We therefore examined the sera of patients with a variety of lymphoreticular neoplasms whose malignant cells might represent lymphoid cells in states of incomplete maturation and/or activation. As can be seen in Fig. 7, the sera of several groups of patients with lymphoreticular malignancies contained elevated levels of IL-2R. Those patients with the human tumor leukemia virus 1 (HTLV-1)-associated adult T-cell leukemia and the Sézary syndrome had the most elevated levels of receptor. Patients with Hodgkin's desease, chronic lymphocytic leukemia, and solid tumors had lower levels, but the mean values for each group were still significantly greater ($p > 0.01$) than normals. Patients with multiple myeloma had values which were not significantly different from controls (data not shown).

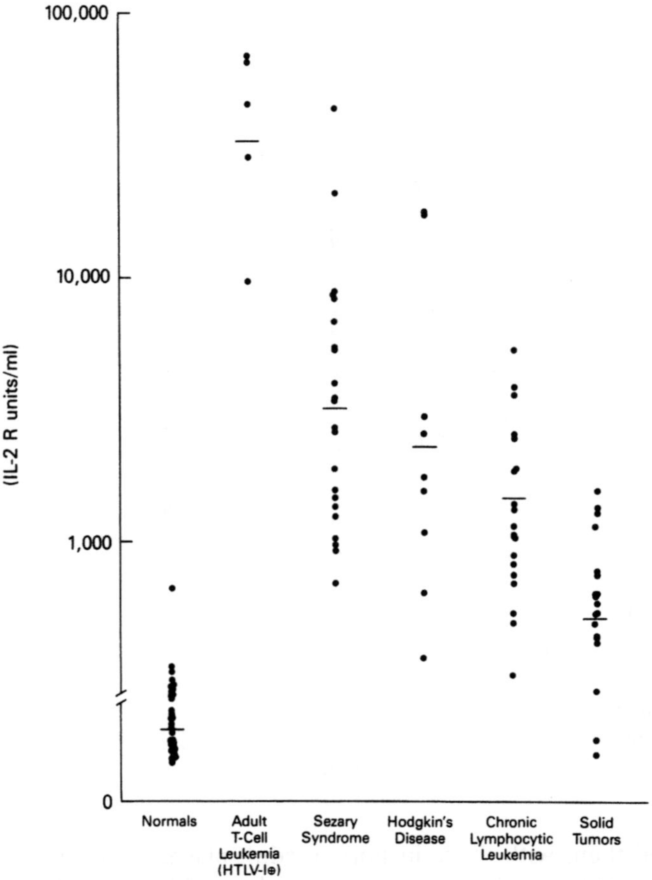

Fig. 7. Serum IL-2R levels in normal controls and patients with various lymphoreticular neoplasms and solid organ cancers. The *horizontal bar* represents the mean value for that group of patients. Results are expressed as units of IL-2R per milliliter of serum

4 Discussion

In the present studies we have examined the capacity of neonatal cord blood cells to respond to IL-2 and to express both cell-associated and soluble IL-2R. Freshly isolated CBMC, like adult PBMC, did not express IL-2R, and the plasma levels of IL-2R from cord blood were comparable to that of adults. Following in vitro activation with the T-cell mitogens PHA and OKT3, CBMC, like adult PBMC, proliferated vigorously and expressed amounts of cell-associated IL-2R comparable to that of adult PBMC. In vitro-activated CBMC also released soluble IL-2R into culture supernatants, generating less supernatant IL-2R with OKT3 stimulation than PBMC but secreting similar amounts with PHA and IL-2 stimulation. CBMC and adult PBMC released IL-2R with similar kinetics. Cultures of CBMC stimulated with PWM secreted IL-2R but not IgM. Studies of the level of serum

IL-2R in patients with various forms of lymphoreticular neoplasms which might represent immature cells in various states of activation revealed elevated levels of IL-2R in several patient groups.

Recent studies have begun to elucidate the critical role IL-2 plays in the maturation and regulation of the immune response. Previous studies of this lymphokine network in neonatal cells had demonstrated normal IL-2 production by PHA-stimulated neonatal T cells (Hayward and Kurnick 1981; Wakasugi et al. 1985). The normal proliferative responses noted with several T-cell mitogens and CBMC suggested that IL-2R expression might be normal (reviewed in Stiehm et al. 1979; Andersson et al. 1981; Hayward 1981; Toivanen et al. 1981). However, since controversy exists as to the presence of IL-2-independent mechanisms of T-cell proliferation (Koretzky et al. 1983), we undertook an analysis of IL-2R expression by CBMC. Previously, it had been suggested that CBMC had a heightened proliferative responsiveness to IL-2 (Hayward and Kurnick 1981) and that, using fluorescence antibody techniques, IL-2R expression might be greater on activated neonatal T cells than on activated adult T cells (Yokoi et al. 1982; Rubin et al. 1984). Our discovery of a soluble released form of the human IL-2R (Rubin et al. 1985b) and recent experiments demonstrating that the released IL-2R can specifically bind to recombinant DNA-produced IL-2 (Rubin et al. manuscript in preparation) suggested that such a released molecule might be a candidate for several of the suppressive factors described in supernatants of activated neonatal T cells (Olding and Oldstone 1976; Olding et al. 1977; Wolf et al. 1977; Williams and Korsmeyer 1978; Abedin and Kirkpatrick 1980; Miyawaki et al. 1981). Thus, we were interested in whether CBMC might secrete levels of IL-2R significantly greater than adult PBMC in vitro. Our findings of similar levels of immunoreactive IL-2R in the supernatants of cultures with most activators suggest that IL-2R may not be responsible for these suppressive effects observed by others. In addition, the biochemical and functional characteristics (glycoprotein, mol. wt. = 40–45 kD, IL-2 binding, released only with activation) of the released IL-2R are significantly different from those of previously described suppressor factors mentioned above from neonatal cells. In a similar way, serum factors have been described which might modulate the fetal-maternal relationship (Murgita et al. 1977; Hellstrom et al. 1969), and we have demonstrated that alloactivation in vitro with adult PBMC leads to the release of soluble IL-2R (Rubin et al. 1985b). Thus, since newborn T cells are competent with regard to alloreactivity, it seemed possible that we might find elevated levels of IL-2R in cord plasma. Again, we found levels of IL-2R comparable to those present in adult plasma. It will be important to document whether or not the immunoreactive material measured in the supernatant and plasma of neonatal and adult cells has the same affinity for IL-2. The observation that CBMC secrete IL-2R but not IgM following activation suggests that these two responses are independently regulated or mature at different times.

Since newborn cells which might be in a state of activation and/or maturation different from adult cells seemed capable of releasing soluble IL-2R, we studied the sera of patients with various lymphoreticular malignancies for the levels of IL-2R. We found significant elevations of IL-2R in certain patients groups but not other (i.e., multiple myeloma – a neoplasm of mature B cells). Since infection of normal T cells with HTLV-1 leads to the release of large amounts of soluble IL-2R (Rubin et al. 1985b), it seems likely that in patients with the HTLV-1-associated adult T-cell leukemia, this serum IL-2R is the result of the tumor releasing IL-2R which could block its rejection by the host's immune response. In certain other patients where the tumors would not be expected to synthesize IL-2R, such as various solid tumors, it is likely that the serum levels of IL-2R reflect the host's response to the tumor. The analysis of serum levels of soluble IL-2R may prove to be of value in the diagnosis and/or monitoring of disorders of immunity in various pediatric patients. The present studies suggest that neonatal cells are fully mature with regard to the IL-2/IL-2R network.

Acknowledgements. M. Elizabeth Fritz and C. C. Kurman provided excellent technical assistance; Dr. R. Yarchoan provided advice regarding the ELISA assays and the software programs for the calculation of the IgM and IL-2R ELISA assays; Drs. W. E. Biddison and L. A. Rubin provided the monoclonal antibody 7G7B6 and the ELISA assay for IL-2R; and Dr. Rubin analyzed the sera from normal patients and those with lymphoreticular malignancies for IL-2R.

References

Abedin M, Kirkpatrick CH (1980) Immunosuppressive activity of cord blood leukocytes. Pediatrics 66:405–410

Andersson U, Bird AG, Britton S, Palacios R (1981) Humoral and cellular immunity in humans studied at the cell level from birth to two years of age. Immunol Rev 57:5–38

Cantrell DA, Smith KA (1984) The interleukin-2 Tcell system: a new cell growth model. Science 224:1312–1316

Hayward AR (1981) Development of responses and interactions in the human fetus and newborn. Immunol Rev 57:39–60

Hayward AR, Kurnick J (1981) Newborn T cell suppression: early appearance, maintenance in culture and lack of growth factor suppression. J Immunol 126:50–53

Hellström KE, Hellström I, Brawn J (1969) Abrogation of cellular immunity to antigenically foreign mouse embryonic cells by a serum factor. Nature 224:914–915

Koretzky GA, Daniele RP, Greene WC, Nowell PC (1983) Evidence for an interleukin-independent pathway for human lymphocyte activation. Proc Natl Acad Sci USA 80:3444–3447

Miyawaki T, Moriya N, Nagaoki T, Kubo M, Yoki T, Taniguchi N (1981) Mode of action of humoral suppressor factor derived from pokeweed mitogen-stimulated cord T-cells on adult B-cell differentiation. J Immunol 126:282–285

Murgita RA, Groidl EA, Kontiainen S, Wigzell H (1977) α Fetoprotein induces suppressor T-cells in vitro. Nature 267:257–259

Nelson DL, Biddison WE, Shaw S (1983) Defective in vitro production of influenza virus specific cytotoxic T-lymphocytes in ataxia-telangiectasia. J Immunol 130:2629–2634

Olding LB, Oldstone MBA (1976) Thymus-derived peripheral lymphocytes from human newborns inhibit division of their mothers' lymphocytes. J Immunol 116:682–686

Olding LB, Murgita RA, Wigzell H (1977) Mitogen stimulated lymphoid cells from human newborns suppress the proliferation of maternal lymphocytes across a cell-impermeable membrane. J Immunol 119:1109–1114

Rubin LA, Kurman CC, Biddison WE, Brown T, Goldman ND, Nelson DL (1984) Studies of the expression of a T-cell activation antigen in primary immunodeficiency disorders and neonatal cord blood cells. In: Griscelli C, Vossen J (eds) Progress in immunodeficiency research and therapy I. Elsevier Science Publishers B V, Amsterdam, pp 103–108

Rubin LA, Kurman CC, Biddison WE, Goldman ND, Nelson DL (1985a) A monoclonal antibody, 7G7/B6, that binds to an epitope on the human IL-2 receptor distinct from that recognized by IL-2 or anti-Tac. Hybridoma 4:91–102

Rubin LA, Kurman CC, Fritz ME, Biddison WE, Boutin B, Yarchoan R, Nelson DL (1985b) Soluble interleukin-2 receptors are released from activated human lymphoid cells in vitro. J Immunol (in press)

Stiehm ER, Winter HS, Bryson YJ (1979) Cellular (T cell) immunity in the human newborn. Pediatrics Suppl, p 814

Toivanen P, Uksila J, Leino A, Lassila O, Hirvonen T, Ruuskanen O (1981) Development of mitogen responding T-cells and natural killer cells in the human fetus. Immunol Rev 57:89–105

Tosato G, Magrath IT, Koski IR, Dooley NJ, Blaese RM (1980) B cell differentiation and immunoregulatory T cell function in human cord blood lymphocytes. J Clin Invest 66:383–388

Uchiyama T, Broder SA, Waldmann TA (1981) A monoclonal antibody reactive with activated and functionally mature human T cells. J Immunol 126:1393–1397

Wakasugi N, Virelizier JL (1985) Defective IFN production in the human neonate. I. Dysregulation rather than intrinsic abnormality. J Immunol 134:167–171

Wakasugi N, Virelizier JL, Arenzana-Seisdedos F, Rothhut B, Mencia-Huerta J-M, Russo-Marie F, Fiers W (1985) Defective IFN-γ production in the human neonate. II. Role of increased sensitivity to the suppressive effects of prostaglandin E. J Immunol 134:172–176

Waldmann TA, Goldman CK, Robb RJ, Depper JM, Leonard WJ, Sharrow SO, Bongiovanni KF, Korsmeyer SJ, Greene WC (1984) Expression of interleukin-2 receptors on activated human B-cells. J Exp Med 160:1450–1466

Williams RC Jr, Korsmeyer SJ (1978) Studies of human lymphocyte interactions with emphasis on soluble suppressive activity. Clin Immunol Immunopathol 9:335–349

Wolf RL, Lomnitzer R, Rabson, AR (1977) An inhibitor of lymphocyte proliferation and lymphokine production released by unstimulated foetal monocytes. Clin Exp Immunol 27:464–468

Yarchoan R, Nelson DL (1983) A study of the functional capabilities of human neonatal lymphocytes for in vitro specific antibody production. J Immunol 131:1222–1228

Yarchoan R, Murphy B, Strober W, Schneider HS, Nelson DL (1981) Specific anti-influenza virus antibody production in vitro by human peripheral blood mononuclear cells. J Immunol 127:2588–2594

Yarchoan R, Biddison WE, Schneider HS, Nelson DL (1982) Human T-cell subset requirements for the production of specific anti-influenza virus antibody in vitro. J Clin Immunol 2:118–125

Yarchoan R, Engler RJM, Barrow LA, Kurman CC, Nelson DL (1984) Antigen stimulated human B cell clones are committed to the production of a single class of antibody in vitro. J Immunol 132:2696–2699

Yokoi T, Miyawaki T, Yachie A, Ohzeki S, Taniguchi N (1982) Discrepancy in expression ability of Tac antigen and Ia determinants defined by monoclonal antibodies on activated or cultured cord blood T-lymphocytes. J Immunol 129:1441–1445

Yu VYH, Waller CA, Maclennan ICM, Baum JD (1975) Lymphocyte reactivity in pregnant women and newborn infants. Br Med J 1:428–432

Development of Lymphocyte Responses to Herpes Simplex Virus Following Neonatal Infection

ANTHONY HAYWARD, PAUL LEIBSON, and ANN ARVIN

Departments of Pediatrics, University of Colorado School of Medicine and Stanford University, Denver, CO 80262, USA

1 Introduction

Children with congenital cytomegalovirus infection continue to shed the virus in their urine for several years, and survivors of neonatal herpes simplex virus (HSV) infection have recurrent crops of skin vesicles caused by HSV in infancy. Both observations suggest that immune responses to these viruses are limited following congenital or neonatal infection. Affected children have few T cells in their blood, which proliferate in response to the infecting virus (Hayward et al. 1984a). However, the presence of first IgM and later IgG antibodies to HSV in their serum indicates that they have developed specific immunity and suggests that they have T cells capable of providing help to B cells (Hayward et al. 1984b).

Mechanisms which might account for the apparent dichotomy between antibody and T-cell mediated HSV-specific responses in the survivors of perinatal HSV infection include: suppression of T-cell proliferative responses by specific suppressor T cells and/or interference with T-cell clonal expansion as a direct or indirect result of infection. In this communication we review first the potential for herpesvirus-specific immune responses by mononuclear cells (MNC) from human newborns and second the immune status of survivors of neonatal infection. New data to be presented suggest that as the subjects recover from their cutaneous HSV recurrences there are increases in both HSV-specific T cells and HSV-directed natural killer (NK) cell-mediated cytotoxicity (Fitzgerald et al. 1985).

2 Clinical Aspects of Neonatal HSV Infections

Infection with HSV usually follows fetal contact with genital lesions during birth. Seventy-five percent of cases are HSV type 2, 25% type 1; and current data on outcome indicate that 30% recover, 60% become severely neurologically impaired, and 10% die. The incidence of neonatal HSV appears to be increasing (Kohl 1984) and in Seattle is now about 18 per 10^5 live births (Corey et al. 1983). The presence of maternally derived serum antibody to HSV may provide partial protection to the infant (Yeager et al. 1980). Treatment with acyclovir has increased the number of survivors, many of whom subsequently experience

Immunology of the Neonate
Burgio, Hanson, Ugazio (Eds.)

cutaneous reactivations of the virus, usually in the form of crops of blisters on the limbs or trunk.

2.1 HSV-Specific Responses by Newborn Cells

2.1.1 HSV-Specific T cells

About 65% of newborn MNC are T cells (identified by monoclonal anti-T3 antibody) and their T4/T8 subset distribution is similar to adults. The newborn cells proliferate in response to mitogens and make IL 2. Their response to alloantigen in mixed lymphocyte culture has the same frequency and phenotype of responders as that of adults (Hayward and Malmberg 1984). The frequency of newborn T cells which respond to microbial antigen is much lower than that of adults: e.g., $< 1{:}10^6$ for HSV and tetanus toxoid (Chilmonczyk et al. 1985). Adults who are nonimmune to CMV or HSV tend to have higher responder cell frequencies (about $1{:}10^5$), possibly because of cross-reactivity with other antigens. The newborn T-cell repertoire is generated through recombination of the V, D, and J segments of the variable regions of the α and β chains of the T-cell receptor (Collins et al. 1985). This process takes place in the thymus and must be linked to the tolerization or elimination of self-reactive cells. Whatever mechanisms are responsible, it seems clear that T-cell clonal diversity is established by birth and that antigen stimulation is responsible for the development of the adult repertoire of memory cells.

2.1.2 Nonspecific Effector Mechanisms

The subset of human NK cells which responds to HSV infected cells is mostly Leu 11^+ and Leu 7^- (Lanier et al. 1983; Lopez et al. 1983); the HSV determinants they recognise are probably virus encoded glycoproteins (Bishop et al. 1984). Leu 11^+ cells comprise 5% of newborn MNC compared with a mean of 12% in adults. Newborn NK cells are less effective at killing classical NK targets than adult cells (Kohl et al. 1984), and our own results show that newborn MNC do not suppress HSV plaque formation in an in vitro assay which requires NK cells (Table 1).

The extent to which the lack of newborn NK activity is due to intrinsic immaturity of the cells or secondary to a lack of lymphokine stimulation is

Table 1. In vitro function of newborn NK cells[a]

	Test system	
NK cells	HSV plaque suppression	K562 lysis
Newborn	2% ± 2%	5% ± 3%
Adult	60% ± 10%	33% ± 8%

[a] Results are means of > 8 samples per group. Methods are given in Leibson et al. (submitted)

Table 2 γ-Interferon production by MNC as units/ml[a]

Donors	*n*	0	IL 2	HSV	Both
Newborn	6	1(1)	3(1)	3(1)	53(2)
-Leu 11		0(1)	1(1)	2(1)	6(1)
Imune adult	5	2(1)	28(5)	106(6)	105(6)
-Leu 11		1(1)	14(3)	32(4)	40(3)

[a] 10^6 cells were incubated for 3 days and then supernatant γ-interferon was measured by radioimmunoassay.
n, number of samples assayed; -Leu 11, MNC depleted by anti-Leu 11 antibody and C; figures in parentheses, ± 1 SE range

unknown (Miyawaki et al. 1985). Our results suggest that preactivation of newborn MNC with IL 2 or α-interferon can sometimes increase their anti-HSV activity to adult levels in the plaque suppression test.

Production of γ-interferon is a recently identified property of NK cells. Our recent results show that newborn MNC can make γ-IFN in response to HSV provided that IL 2 is also supplied (Table 2). Our tests of newborn NK cells therefore suggest that they are capable of increased levels of function provided that soluble mediators are supplied. Since T cells are a major source of mediators, such as IL 2, neither effector mechanisms can be considered in isolation.

3 Response to HSV Infection in Newborns

3.1 T Cell Responses

Our own results (Table 3) and those of Pass et al. (1981) suggest that infants who were neonatally infected with HSV have reduced HSV responder T-cell frequencies (HSV-RCF) in their blood for months or years after they have recovered from the primary infection.

Defective monocyte function (Stiehm et al. 1984), seems an unlikely explanation for the reduced HSV-RCF as the patients' monocytes support normal

Table 3 Responses of lymphocytes from HSV survivors

Subjects	PHA response	MLC response	HSV RCF
Neonatal HSV	N	N	1: 50 000
Age matched non-immune children	N	N	< 1:100 000
Age matched immune children	N	N	1: 10 000

N, normal; HSV RCF, HSV specific responder. T cell frequency determined by limiting dilution. (For methods see Hayward et al. 1984b.)

Table 4. Presentation of antigen to maternal T-cell blasts by adherent cells from infants following neonatal HSV infection[a]

Cultures with:	0	Ag	Monoc	Ag+monoc
Thymidine uptake:	250	1 150	2 200	13 740

[a] Maternal HSV-specific T-cell blasts were recultured for 40 h with HSV antigen and/or child-adherent cells and then pulsed with thymidine. (For method see Hayward et al. 1984c.)

mitogen responses and process and present antigen effectively to maternal HSV-specific T-cell blasts (Table 4). Nonspecific immunosuppression is also unlikely as the patients' MNC responded normally to mitogen and alloantigen stimulation. The culture conditions used for limiting dilution stimulate predominantly T 4^{+} cells (Suez et al. 1985), and it is possible to deplete T 8^{+} cells by panning before measuring the RCF. Under these circumstances, the HSV-RCF does not rise significantly; so it seems unlikely that suppression by T 8^{+} cells is responsible for the low HSV-RCF.

Our results do not exclude the existence of T 4^{+} suppressor cells, but available methods do not permit us to test for these. The presence of anti-HSV antibodies in the patients' sera suggests that they have at least some HSV-specific T cells in their lymphoid tissue, even if these cannot be detected in a proliferation assay. To test for nonproliferating T cells in the patients' blood, we measured IL 2 production by their MNC in the presence and absence of HSV. The results (Table 5) indicate that small amounts of IL 2 are made even by MNC from infants who are still experiencing cutaneous HSV recurrences, though the levels are less than those seen with adult immune cells.

3.2 NK Cells Following Neonatal Infection

The percentage of Leu 11^{+} cells in the blood of four children with cutaneous recurrences of HSV following neonatal infection was normal (range 6%–12%) in each. Infants with congenital CMV have increased NK cell activity (Harrison and Waner 1985), but there are no reports of NK function following neonatal HSV. We therefore measured the ability of patients' MNC to suppress HSV replication

Table 5. In vitro IL 2 production by HSV stimulated MNC[a]

MNC source	Arbitrary IL 2 units
Symptomatic patients	50
Recovered patients	100
Healthy adults	300

[a] 10^{6} MNC were irradiated with 2000 rads and cultured with HSV or control antigen for 3 days. IL 2 activity in the supernatant was determined by dilution on an IL 2-dependent cell line. Control backgrounds were subtracted and the HSV-induced IL 2 production is expressed as the greatest dilution sustaining cell proliferation

Table 6. Recent findings in *HSV* survivors

Symptoms	Age	% Leu 11+	HSV-RCF	HSV suppn
0	7 years	12	1: 8 000	80%
0	7 months	11	1:12 000	71%
+	9 months	6	1:60 000	0%
+	6 months	7	1:60 000	0%

+, with cutaneous recurrences; 0, no more cutaneous recurrences; HSV-RCF, responder T-cell frequency measured by limiting dilution; HSV suppn, percentage suppression of HSV plaques in a fibroblast monolayer

in the plaque suppression assay. The results in a small number of tests (Table 6) suggest that suppression of HSV replication in the plaque suppression assay is low during the time that the patients are still having cutaneous recurrences and that it increases as the T-cell RCF rises and the cutaneous recurrences cease.

4 Correlations with Animal Models

The available data suggest that antibody protects against HSV by neutralization and by preventing the infection of susceptible cells, as shown with high specificity by idiotype suppression (Kennedy et al. 1984). Antibody may continue to act even after nerve tissue has become infected (Simmons and Nash 1985), by preventing the virus from returning to the skin from sensory nerves.

Virus within cells is not accessible to antibody, and cellular responses are required to limit replication and, ultimately, to eliminate the virus. NK cells, which are activated and proliferate during the course of virus infections (Biron and Welsh 1985), appear to be the principal effectors, and interference with their function increases the severity of herpesvirus infection (e.g., steroids and CMV infection; Quinnan and Manischiewitz 1978). Recently in vivo transfer of as few as 3×10^5 cloned NK cells per mouse was shown sufficient to protect mice from lethal HSV infection (Bukowski et al. 1985). The small number of cells required is perhaps attributable to their activated state. In the related model described by Kohl et al. (1984), adult (but not newborn) human MNC were able to protect newborn mice from lethal HSV only when they were transferred along with anti-HSV antibody and α-interferon.

Lymphokines, predominantly IL-2 (Henney et al. 1981; Trinchieri et al. 1984), α-interferon (Herberman et al. 1979), and γ-interferon are all known to activate NK cells and to amplify their anti-herpesvirus effects. The T-cell contribution to HSV immunity seems likely to depend as much on the production of soluble mediators as on direct cytotoxicity, and this may account for the confusion in the literature over the relative importance of MHC class-I and class-II restricted T cells for antivirus responses. In the case of herpesviruses (as distinct from influenza or LCM viruses), it appears that class-II restricted cells are sufficient for the clearance of virus from the ear of acutely infected mice (Larsen et al. 1984) and for the transfer of delayed-type hypersensitivity (Nash et al. 1981).

5 Conclusions

Factors which may increase human newborns' susceptiblility to severe infections include a deficiency of functional NK cells (Abo et al. 1982; Kohl et al. 1981), reduced binding of target cells to NK cells (Bailey and Schacter 1985) less α- (Nair et al. 1985) and γ-interferon (Wilson et al. 1985) production. Although these defects have been thought of as primary our data suggest that some e.g., γ-IFN production) are secondary to a lack of T-cell derived lymphokines. This in turn reflects the low frequency of T cells with specificity for HSV in newborns (Chilmoczyk et al. 1985).

HSV infections after the first weeks of life result in the appearance of HSV-specific responder T cells in the blood with a frequency identical to that found in adults. Neonatal infection, in contrast, is severe and may be followed by continued viral activity without eliciting a comparable expansion of the HSV responder cell pool. The reasons for this difference are unknown but they do not appear to include defects of antigen processing (Hoffman et al. 1981).

Interference with antivirus immunity as a result of virus infection is well documented (Sullivan et al. 1975) and conventionally attributed to effects on monocytes or to suppressor cell induction (Reinherz et al. 1980). Selective interference by viruses with highly specialized cell functions, such as growth hormone production in the absence of cytotoxicity, have been identified more recently (Oldstone et al. 1984). HSV is known to infect proliferating T cells (Rinaldi 1978), so direct infection of responder cells is one possible explanation for the lack of clonal expansion.

References

Abo T, Cooper MD, Balch CM (1982) Postnatal expansion of the natural killer and killer cell population in humans identified by the monoclonal HNK 1 antibody. J Exp Med 155:321–326

Bailey J, Schacter B (1985) Stimulation of cord and adult NK cells. Pediatr Res (in press)

Biron CA, Welsh RM (1985) Blastogenesis of natural killer cell in vivo results in progeny cells with the capacity for continued proliferation. Fed Proc 44:1314 (Abstract # 5282)

Bishop GA, Marlin SC, Schwartz SA, Glorioso JC (1984) Human natural killer cell recognition of herpes simplex virus type I glycoproteins: specificity analysis with the use of monoclonal antibodies and antigenic variants. J Immunol 133:2206–2209

Bukowski JF, Warner JF, Dennert G, Welsh RM (1985) Adoptive transfer studies demonstrating the anti-viral effects of natural killer cells in vivo. J Exp Med 161:40–52

Chilmonczyk BA, Levon MJ, McDuffi R, Hayward AR (1985) Characterization of the human newborn response to herpesvirus antigen. J Immunol 134:4184–4188

Collins MKL, Goodfellow PN, Spurr NK, Solomon E, Tanigawa T, Tonegawa S, Owen MJ (1985) The human T cell receptor β-chain gene maps to chromosome 14. Nature 314:273–274

Corey L, Adams HG, Brown ZA, Holmes KK (1983) Genital herpes simplex infections: clinical manifestations, course and complications. Ann Intern 998:958–972

Fitzgerald PA, Mendelsohn M, Lopez C (1985) Human natural killer cells limit replication of herpes simplex virus type 1 in vitro. J Immunol 134:2666–2672

Harrison CJ, Waner JL (1985) Natural killer cell activity in infants and children excreting cytomegalovirus. J Infect Dis 151:301–307

Hayward AR, Malmberg S (1984) Response of human newborn lymphocytes to alloantigen: lack of evidence for suppression induction. Pediatr Res 18:414–419

Hayward AR, Herberger M, Levin M (1984a) Reduced cytomegalovirus (CMV) responder T cell frequency in congenital CMV infection. In: Rapp F (ed) Herpesvirus. Liss, New York p 211
Hayward AR, Herberger MJ, Groothuis J, Levin MJ (1984b) Specific immunity after congenital or neonatal infection with cytomegalovirus or herpes simplex virus. J Immunol 133:2469–2473
Hayward AR, Lazslo M, Levin M (1984c) Suppression of HSV proliferation in fibroblast monolayers by HLA matched and unmatched mononuclear cells. Proc. 16th Leukocyte Culture Conference
Henney CS, Kuribayashi K, Kern DE, Gillis S (1981) Interleukin 2 augments natural killer cell activity. Nature 291:335–338
Herberman RB, Ortaldo JR, Barnard G (1979) Augmentation by interferon of human natural and antibody dependent cell mediated cytotoxicity. Nature 277:221–223
Hoffman AA, Hayward AR, Kurnick JT, DeFreitas E, McGregor J, Harbeck RJ (1981) Presentation of antigen by human newborn monocytes to maternal tetanus toxoid specific T cell blasts. J Clin Immunol 1:217
Kennedy RC, Adlerstorthz K, Burns JW, Henkel RD, Dressman GR (1984) Antiidiotype modulation of herpes simplex virus infection leading to increased pathogenicity. J Virol 50:951–953
Kohl S (1984) Neonatal herpes. In: Ogra PL (ed) Neonatal infections. Grune and Stratton, New York
Kohl S, Bigelow RH, Loo LS (1984) Ontogeny of protection of neonatal mice from lethal herpes simplex virus infection by human leukocytes, antiviral antibody and recombinant α interferon. Pediatr Res 18:1164–1167
Kohl S, Loo LS, Gonik B (1984) Analysis in human neonates of defective antibody-dependent cellular cytotoxicity and natural killer cytotoxicity to herpes simplex. J Infect Dis 150:14
Lanier LL, Le AM, Phillips JH, Warner NL, Babcock GF (1983) Sub-populations of human natural killer cells defined by expression of the Leu-7 (HNK-1) and Leu-11 (NK-15) antigens. J Immunol 131:1789–1796
Larsen HS, Feng MF, Horohov DW, Moore RN, Rouse BT (1984) Role of T-lymphocyte subsets in recovery from herpes simplex virus infection. J Virol 50:56–59
Lopez C, Kirkpatrick D, Read SE, Fitzgerald PA, Pitt J, Pahwa S, Ching CY, Smithwick EM (1983) Correlation between low natural killing of fibroblasts infected with herpes simplex virus type I and susceptibility to herpes virus infections. J Infect Dis 147:1030–1035
Miyawaki T, Seki H, Taga K, Sato H, Taniguchi N (1985) Dissociated production of interleukin-2 and immune (γ) interferon by phytohaemagglutinin-stimulated lymphocytes in healthy infants. Clin Exp Immunol 59:505–511
Nair MPN, Schwartz SA, Menon M (1985) Decreased natural and antibody-dependent cellular cytotoxicity is associated with decreased production of natural killer cytotoxic factors and interferon in neonates. Pediatr Res 19:279A (# 1011)
Nash AA, Phelan J, Woldy P (1981) Cell-mediated immunity in herpes simplex virus infected mice: H2 mapping of the delayed type hypersensitivity response and the antiviral T cell response. J Immunol 126:1260–1262
Oldstone MBA, Rodriquez M, Daughaday WH, Lampert PW (1984) Viral perturbation of endocrine function: disordered cell function leads to disturbed homeostasis and disease. Nature 307:278–280
Pass RF, Dworsky ME, Whithley RJ, August AM, Stagno S, Alford CA (1981) Specific lymphocyte blastogenic responses in children with cytomegalovirus and herpes simplex virus infections acquired early in infancy. Infect Immunol 34:166–170
Quinnan GV, Manischiewitz JE (1978) The role of natural killer cells and antibody-dependent cell mediated cytotoxicity during murine cytomegalovirus infection. J Exp Med 150:1549
Reinherz EL, O'Brien C, Rosenthal P, Schlosssman SF (1980) The cellular basis for viral induced immunodeficiency: analysis by monoclonal antibodies. J Immunol 125:1269–1274
Rinaldo CR, Richter BS, Black PH, Callery R, Chess C, Hirsch MS (1978) Replication of herpes simplex virus and cytomegalovirus in human leukocytes. J Immunol 120:130–136
Simmons A, Nash AA (1985) Role of antibody in primary and recurrent herpes simplex virus infection. J Virol 53:944–948
Stiehm ER, Sztein MB, Steeg PS, Mann D, Newland C, Blaese M, Oppenheim JJ (1984) Deficient DR antigen expression of human cord blood monocytes: reversal with lymphokines. Clin Immunol Immunopathol 30:430–436
Suez D, Hayward AR (1985) Phenotyping of proliferating cells in cultures of human lymphocytes. J Immunol Methods 78:49–57

Sullivan JL, Barry DW, Alabrecht P, Lucas SJ (1975) Inhibition of lymphocyte stimulation by measles virus. J Immunol 114:1458–1462
Trinchieri G, Matsumoto-Kobayashi M, Clark SC, Seehra S, London L, Perussia B (1984) Response of resting human peripheral blood natural killer cells to interleukin 2. J Exp Med 160:1147–1169
Wilson CB, Westall J (1985) Lymphokine-dependent macrophage activation in newborns. Pediatr Res 19:284(a)
Yeager AS, Arvin AM, Urbani LE, Kemp JA (1980) Relationship of antibody to outcome in neonatal herpes simplex infections. Infect Immun 29:532–538

T and NK Lymphocyte Subpopulations in the Neonate

RITA MACCARIO and G. R. BURGIO
Department of Pediatrics, University of Pavia, Pavia, Italy

1 Introduction

The peculiar features of T-cell subsets and T-cell functions in the neonatal period probably represent a biological compromise between different interacting and often conflicting factors. The fetal immune system must cope with potentially aggressive maternal cells as well as invading pathogens but at the same time must avoid the risks of the potentially almost as catastrophic effects of a strong reaction against maternal antigens. Obviously, the capacity of eliminating maternal immunocompetent cells or microorganisms eventually entering the fetal circulation would favor an early and complete maturation of T cells which, however, would increase the risks of a strong fetal reaction against maternal antigens.

As for antibody-mediated immunity, the immunoglobulin-producing machinery is already functional around the 8th to 10th week of gestation, so that a potentially analogous biological problem has been solved by the "trick" of "switching off" fetal antibody production, which is substituted by a passive transfer of maternal antibodies. This ensures that the neonate reaches a reasonable level of protection against invading pathogens without endangering the mother. Obviously this trick cannot be used for T lymphocytes, so the neonatal picture of T-cell subsets and functions is characterized by a complicated "jigsaw puzzle" of subsets at various stages of maturation interacting in such a way as to provide "sufficient" protective immunity and at the same time avoiding the risks of dangerous fetomaternal interactions.

2 Neonatal Distribution of T Cell Subsets

Early studies of neonatal T lymphocytes with E rosettes as the only membrane marker gave contradictory results, with some groups reporting percentages of T cell similar to those in adults and others reporting low percentages (Asma et al. 1977; Ugazio et al. 1976). Although the reasons for this discrepancy were obscure, at that time, several groups speculated that the low percentage of T lymphocytes might have resulted from incomplete maturation or intense traffic of immature post-thymic precursors, ultimately contributing to neonatal susceptibility to viral and fungal infections. These early reports also suggested that the study of circu-

Immunology of the Neonate
Burgio, Hanson, Ugazio (Eds.)

lating neonatal lymphocytes might contribute to the understanding of T-cell differentiation, and this has been confirmed recently using monoclonal antibodies (MoAb) and various cell membrane markers.

As outlined in Fig. 1, membrane phenotype determined by means of MoAb and other cellular markers has shown that the distribution of the various T-cell subsets has several characteristic features in the neonate. In particular, the use of pan-T reagents like the OKT3 MoAb has shown that the percentage of T cells is apparently lower in the neonate than in adults (Foa et al. 1984; Maccario et al. 1983; Griffiths-Chu et al. 1984; Solinger 1985). Furthermore, the percentage of E-rosette-forming T cells (E-RFC) is also lower in the neonate but reaches adult values after various treatments known to increase the binding capacity of sheep red blood cells (SRBC), such as pretreatment with neuroaminidase or AET (Porta et al. 1985). Incidentally, this finding shows that previously contradictory reports concerning the percentage of neonatal T cells could be the result of differences in the techniques used by the different groups. In fact, neonatal lymphocytes include a subset which in older individuals can be found only in subjects with T-cell defects, characterized by the expression of membrane receptors with low avidity for SRBC. Furthermore, there is no difference between the percentages of adult and neonatal lymphocytes staining with the cytochemical T-cell marker alpha naphthylacetate esterase (ANAE) (Porta et al. 1985).

The response of neonatal lymphocytes to mitogens such as PHA, which are known to stimulate the proliferation of T-cells irrespective of their subset, does not differ substantially from that of adults (Ugazio et al. 1976). Following activation with T-cell mitogens, a remarkable percentage of adult lymphocytes express HLA DR antigens, which in contrast are expressed only by a minority of activated neonatal T cells (Ko et al. 1979; Miyawaki et al. 1982). However, expression of

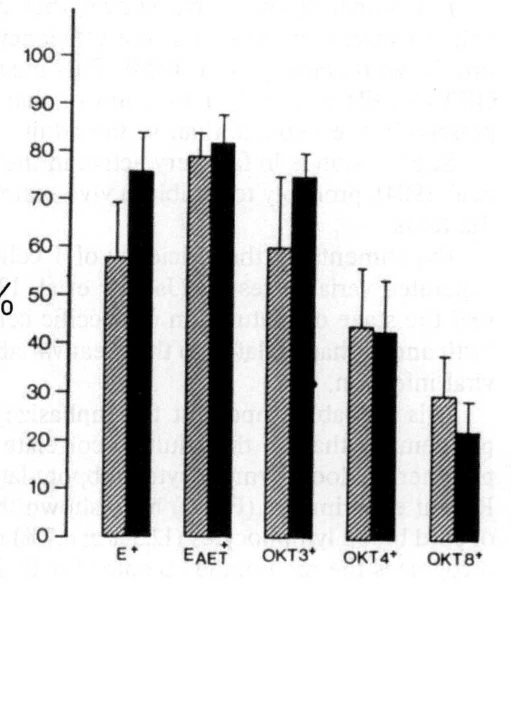

Fig. 1. T lymphocyte subpopulations in adult peripheral blood (■) and in cord blood (▨)

Tac antigen and production of interleukin-2 (IL-2) in response to T-cell mitogens are comparable (Yokoi et al. 1982).

On the whole, these data suggest that there are no substantial differences in the percentages and overall function of T lymphocytes in adults and neonates; at the same time they indicate a significant difference in the distribution of the T-cell subpopulations.

In fact, the study of the T-cell subsets OKT4+ and OKT8+ has shown that while the percentage of OKT4+ cells is similar in neonates and adults, the proportion of OKT8+ cells is significantly higher in the neonate (Maccario et al. 1983). As for the OKT8+ cells, our findings are apparently in contrast with those of Hayward et al. (1981), who reported a lower percentage of OKT8+ cells in the neonate. The discrepancy is only apparent, however, because Hayward et al. evaluated OKT+ cells on T-cell-enriched fractions, and we have shown that neonatal OKT3- cells include a sizeable percentage of OKT8+ lymphocytes (Maccario et al. 1983). This is not the case for adult PBL, where all OKT8+ cells are also OKT3+. Therefore, as discussed later, circulating lymphocytes of the neonate include a subset of OKT3–, OKT4–, OKT8+ cells (Maccario et al. 1983; Vitiello et al. 1984).

Some authors have observed the presence in cord blood of lymphocytes (CBL) expressing OKT4 and OKT8 antigens simultaneously; the values range form 12% ± 12% (Solinger et al. 1985) to 25% ± 17% (Griffiths-Chu et al. 1984). Others have found a small number of cells expressing both antigens only occasionally (Foa et al. 1984). The presence in cord blood of a subpopulation of OKT3–, OKT8+ cells, and of variable percentages of lymphocytes simultaneously expressing OKT4 and OKT8 antigens is in agreement with the observation that in the neonate the sum of the percentages of lymphocytes reacting with OKT4 and OKT8 MoAb is always greater than the percentages of OKT3+ lymphocytes and of E-RFC (Foa et al. 1985; Maccario et al. 1983).

Functional studies have shown that neonatal T lymphocytes have a low helper capacity, measured as the efficiency to induce immunoglobulin production in vitro (Jacoby et al. 1984). This means either that the neonate has fewer OKT4+ cells with helper function or that helper activity is more actively suppressed in the neonate than in the adult.

Suppression is in fact very active in the neonate (Hayward et al. 1981; Jacoby et al. 1984), probably to inhibit in vivo maternal cell proliferation and rejection of the fetus.

Experiments on the efficiency of T cells with specific cytotoxic activity have generated variable results (Jacoby et al. 1984; Toivanen et al. 1981), suggesting that the stage of maturation of specific cell-mediated lympholysis is variable at birth and perhaps related to the great variability in the neonate's susceptibility to viral infection.

It is probably important to emphasize that in the neonate it is even more problematic than in the adult to correlate the presence of surface antigens on peripheral blood lymphocyte subpopulations with their functional activity. Recent experiments (Fig. 2) have shown that a variable but sizeable percentage of cord blood lymphocytes (12.3% ± 6.7%) reacts with the MoAb anti-Tac, which recognizes the membrane receptor for IL-2; in contrast, Tac+ cells are virtually

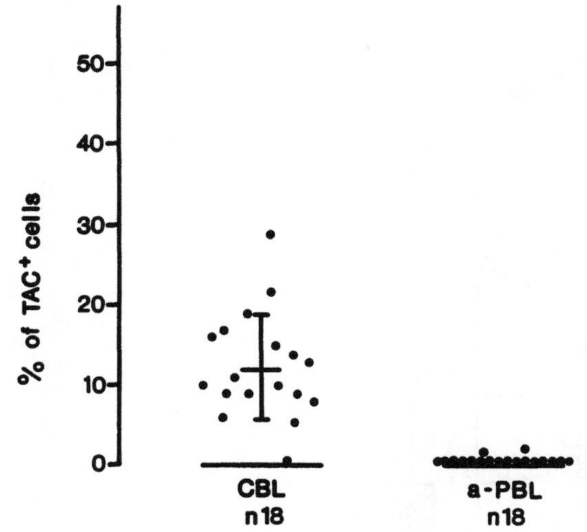

Fig. 2. Proportion of TAC+ cells in human cord blood *(CBL)* and in adult peripheral blood *(a-PBL)*

absent among unstimulated adult peripheral blood lymphocytes (a-PBL) (0.3% ± 0.7%). As Tac is present on a subset of thymocytes as well as on the membrane of activated lymphocytes, neonatal Tac+ cells might represent circulating immature T cells or T cells activated possibly by maternal antigens. Although other explanations for this finding are also conceivable, the expression of the IL-2 receptor might relate to a well known and so far obscure feature of neonatal lymphocytes, i.e., their capacity to proliferate spontaneously in vitro in the absence of mitogenic stimuli.

3 Immature T Cells

Various studies have shown that human cord blood includes lymphocyte subsets characterized by an "immature" or "immature-like" phenotype (Foa et al. 1984; Maccario et al. 1983; Griffiths-Chu et al. 1984; Solinger et al. 1985). Figure 3 shows our results obtained using MoAbs and other cell markers associated with immature and/or activated cells. The percentages of cells bearing the receptor for peanut agglutinin (PNA) is higher in CBL than in a-PBL, and PNA+ cells include both T and non-T lymphocytes (Maccario et al. 1981). The percentages of OKT6+ lymphocytes, not statistically different in newborns and adults in our hands (Maccario et al. 1983), have been reported by others to be definitely high in the neonate (Foa et al. 1984; Griffiths-Chu et al. 1984). This discrepancy may be related to the use of a cell sorter or, in any case, by the "weak" staining obtained with the OKT6 MoAb.

Earlier markers of T-cell maturation, such as nuclear synthesis of terminal-deoxynucleotidyltransferase (Tdt), and membrane expression of the transferrin receptor (OKT9+ cells), have been studied only by a few groups in a limited number of newborns. CBL are apparently TdT-negative in most instances, although exceptionally neonates possess a high percentage of these cells (Foa et

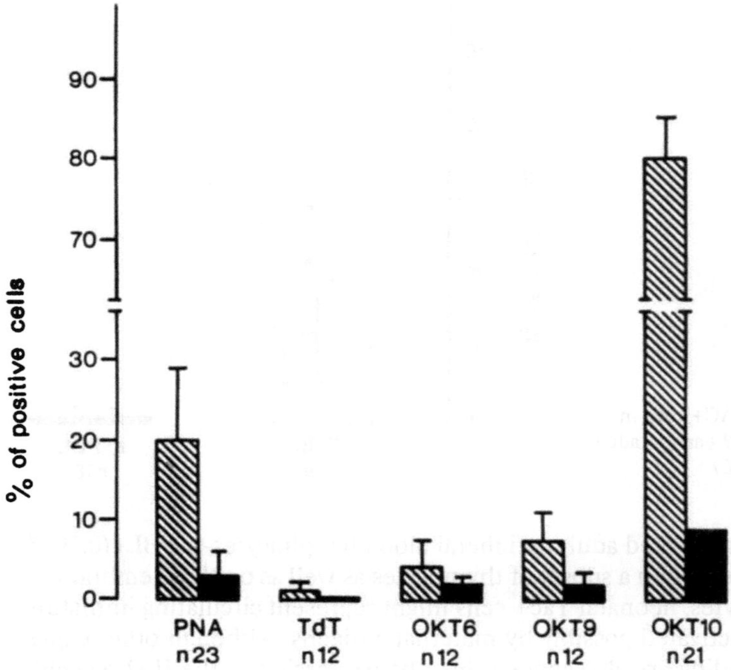

Fig. 3. Immature lymphocyte subpopulations in cord blood (▧) and adult peripheral blood (■)

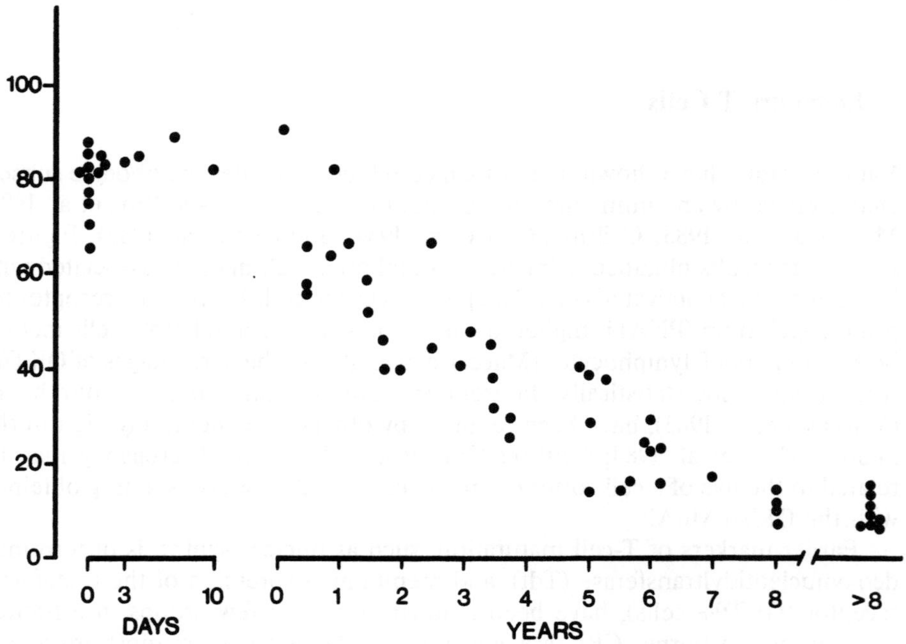

Fig. 4. Distribution of the percentage of OKT10+ lymphocytes at different ages

al. 1984). OKT9+ cells, virtually absent among a-PBL, are apparently present, although in low percentages, in cord blood.

Most CBL are OKT10+ according to several reports (Foa et al. 1984; Griffiths-Chu et al. 1984; Solinger et al. 1984). OKT10+ cells include both T and non-T lymphocytes, and their percentage remains very high during the first 2 years of life (Fig. 4), declining gradually thereafter and not reaching the very low adult values until the age of 6–8 years.

The meaning of these sometimes contradictory data is far from settled and they must be considered with great caution. Several markers of immaturity, such as expression of OKT9 or OKT10, are also markers of activation appearing on the membrane of antigen- or mitogen-stimulated lymphocytes. On the other hand, two factors may condition the functional pattern of cellular subsets in the neonate. Fetomaternal interactions require a particular immunological set-up both on the part of the mother and of the neonate. It is conceivable teleologically that a relative immaturity of neonatal lymphocytes may contribute to the avoidance of potentially catastrophic immunologic interactions between mother and fetus. Alternatively, as delivery is known to result in maternofetal cell transfer, it is equally conceivable that some neonatal phenotypic features, such as expression of OKT9 or OKT10, represent activation by maternal antigens rather than immaturity. The second set of factors probably contributing to the presence of immature cells among CBL is probably related to the intense traffic of immature lymphocytes from primary lymphoid organs (in particular the thymus) to the periphery. In fact, some of the immature neonatal subsets might represent post-thymic precursors, i.e., lymphocytes of the T-cell lineage, partially matured under the influence of the thymus and exported to the periphery where they undergo final maturation (Stutman 1975, 1977, 1978, 1985). Indeed, several of the markers reported are expressed by thymocytes, but not by PBL, after the neonatal period; these include the PNA receptor, expression of the OKT9 antigen, and simultaneous expression by some neonatal lymphocytes of the OKT8 and OKT4 antigens. Of course, as mentioned previously, some of these markers might in fact represent activation rather than immaturity.

The hypothesis that a sizeable proportion of circulating lymphocytes in the neonate represents cells migrating from primary to secondary lymphoid organs also accounts for the great variability observed in the percentages of the various subsets in each study as well as for some of the apparent discrepancies between various studies. In fact, even the absolute number of circulating lymphocytes, which is very high in the neonatal period, shows a high degree of individual and interindividual variability.

The clinical relevance of these findings is obvious. To begin with, the normal neonatal data must be used as a reference for the diagnosis of immunodeficiencies. Furthermore, incomplete maturation of T cells may well underlay, at least in part, the well-known susceptibility of neonates to viral and fungal infections.

4 T and NK Cells: Overlapping Subsets?

NK activity of neonatal lymphocytes against various target cells is low compared with adult values (Table 1), and this has been related to the finding that CBL include a very low percentage of cells reacting with the MoAb HNK-1 (Leu 7) which recognizes most mature NK cells among a-PBL (Vitiello et al. 1984; Abo et al. 1983). As the important role of NK cells in antiviral defense mechanisms has been increasingly recognized, this finding, together with the presence of immature T cells, has also been related to the high susceptibility of the neonate to viral infections.

However, there is no statistical difference between adults and neonates in the percentage of NK cells reacting with NK membrane markers, except HNK-1 (Fig. 5), in particular with MoAb recognizing the Fc receptor of these NK cells. Furthermore, in the neonate, NK activity is highly susceptible to the boosting effect of interferon, and NK activity generated in mixed lymphocyte cultures (NK-like activity) (Table 1) as well as natural cytotoxic activity generated by incubation with IL-2 (lymphokine-activated killer activity, LAK) are similar in adults and newborns (Table 1).

The functional adequacy of neonatal cells with Fc receptor also results from the fact that neonatal cells display ADCC activity comparable to that of adult cells (Table 1). On the whole, these data suggest that the "natural cytotoxic" activity of the neonate is probably adequate, at least under the conditions used in standard laboratory tests with the exception of the spontaneous NK activity. This suggests that the distribution of NK subsets in the neonate differs from that of the adult and that spontaneous NK activity is associated with the functional activity of a subset expressing the HNK-1 antigen. Other normal cytotoxic activi-

Table 1. Characterization of natural cytotoxicity produced by CBL and by *a-PBL*

		K562	DAUDI	HL-60	P815
NK	CBL	2.7 ± 0.9	< 1	< 1	< 1
n 32	*a-PBL*	*11.5* ± 3.5	< 1	< 1	< 1
ADCC	CBL	ND	ND	ND	21.1 ± 6.7
n 22	a-PBL	ND	ND	ND	23.2 ± 3.2
NK-LIKE	CBL	17.0 ± 5.5	< 1	18.4 ± 7.0	< 1
N 22	a-PBL	19.5 ± 2.8	< 1	19.2 ± 6.4	< 1
LAK	CBL	43.2 ± 10.1	30.4 ± 8.6	ND	< 1
N 22	a-PBL	33.3 ± 6.8	21.6 ± 2.5	ND	< 1

ND, not done

Result were quantitated at various effector-to-target cell ratios by calculating number of lytic units per 10^6/cells. One lytic unit is the number of effector cells necessary to lyse 30% of the target cells during the assay period

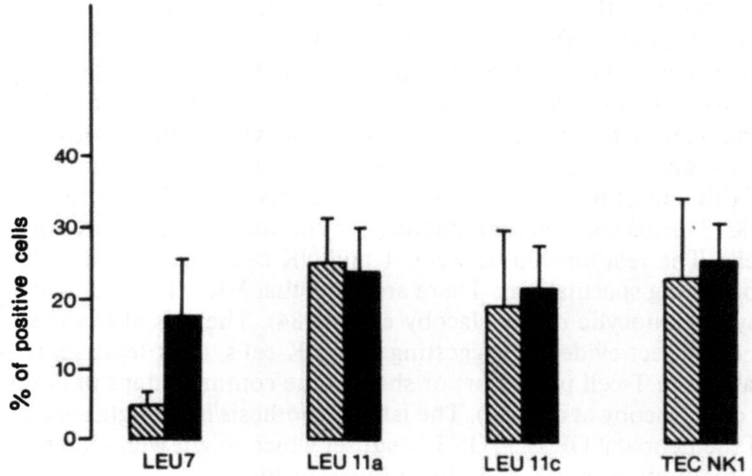

Fig. 5. Proportion of Leu7+ (HNK-1), Leu 11a+ (NK15) Leu 11c+ (B73.1) and TECNK1+ (AB8.28) cells in human cord blood (▨) and adult peripheral blood (■) mononuclear cells

ties may be related to the presence of NK or NK-like subsets expressing the Fc receptor of NK cells, but not the HNK-1 antigen.

As previously mentioned, about one-third of neonatal OKT3-, non-T cells are also OKT8+ (Maccario et al. 1983). This unusual phenotype has been associated

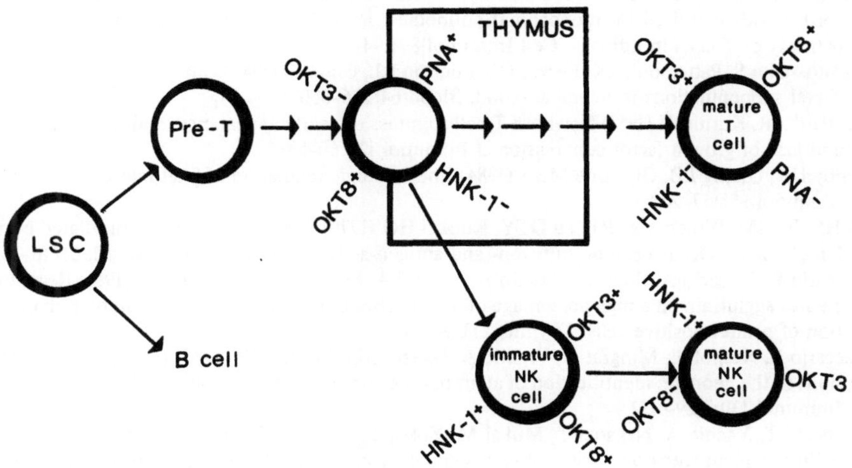

Fig. 6. Hypothetical scheme of the interactions between T-cell and NK-cell ontogeny. According to the hypothesis, the neonatal subset of OKT3-, OKT8+ lymphocytes represents an intermediate stage of differentiation between the pre-T lymphocytes and thymocytes. These cells are able either to differentiate in the thymus along the T cell axis or to leave the thymus and, reaching secondary lymphoid compartments as OKT3+, OKT8+, HNK-1+ cells (Abo et al. 1983), differentiate into mature NK cells. The presence of a high percentage of OKT3-, OKT8+ cells in neonatal peripheral blood could be related to the high rate of recirculating immature cells in the neonatal period

with NK functional activity susceptible to the boosting effect of interferon. This cell subset does not express either the HNK-1 or the OKM1 antigens present on most mature NK cells of the adult, but includes a high percentage of cells expressing the Fc receptor of NK cells, i.e., reactive with the B73.1 (Leu 11c) MoAb. This subset proliferates spontaneously in vitro and in response to PHA and ConA and produces interleukin-1 (IL-1) as well as IL-2. Natural cytotoxic activity of this subset induced by IL-2 (LAK) is very high. Therefore, neonatal OKT8+, OKT3- cells share several phenotypic and functional features of both T and NK cells. The relationship between T and NK cells has been the object of different conflicting speculations. There are hints that NK cells pertain to a cell lineage of myelomonocytic origin (Jacoby et al. 1984). There is also a great deal of direct and indirect evidence suggesting that NK cells, or at least some of them, originate from T-cell precursors or share some common steps of differentiation with T cells (Jacoby et al. 1984). The latter hypothesis is strengthened by identification of the neonatal OKT8+, OKT3- subset which might well represent an early stage of maturation of NK cells, as outlined in Fig. 6.

References

Abo T, Miller CA, Gartland L, Balch CM (1983) Differentiation stages of human natural killer cells in lymphoid tissue from fetal to adult life. J Exp Med 157:273–284

Asma GEM, Schmit HRE, Hijmans W (1977) The determination of numbers of T and B lymphocytes in the blood of children and adults by the direct immunofluorescence technique. Clin Exp Immunol 29:286–291

Foa R, Giubellino MC, Fierro MT, Lusso P, Ferrando ML (1984) Immature T lymphocytes in human cord blood identified by monoclonal antibodies: a model for the study of the differentiation pathway of T cells in humans. Cell Immunol 89:194–201

Griffiths-Chu S, Patterson JAK, Berger CL, Edelson RL, Chu AC (1984) Characterization of immature T cell subpopulations in neonatal blood. Blood 64:296–300

Hayward AR, Kurnick J (1981) Newborn T cell suppression: early appearance, maintenance in culture, and lack of growth factor suppression. J Immunol 126:50–53

Jacoby DR, Olding LB, Oldstone MBA (1984) Immunologic regulation of fetal-maternal balance. Adv Immunol 35:157–208

Ko HS, Fu SM, Winchester RJ, Yu DTY, Kunkel HG (1979) Ia determinants on stimulated human T lymphocytes. Occurrence on mitogen- and antigen-activated T cells. J Exp Med 150:246–255

Maccario R, Ferrari FA, Siena S, Vitiello A, Martini A, Siccardi AG, Ugazio AG (1981) Receptors for peanut agglutinin on a high percentage of human cord blood lymphocytes: phenotype characterization of peanut-positive cells. Thymus 2:329–337

Maccario R, Nespoli L, Mingrat G, Vitiello A, Ugazio AG, Burgio GR (1983) Lymphocyte subpopulations in the neonate: identification of an immature subset of OKT8-positive, OKT3-negative cells. J Immunol 130:1129–1131

Miyawaki T, Yachie A, Nagaoki T, Mukai M, Yokoi T, Uwadana N, Taniguchi N (1982) Expression ability of Ia antigens on T cell subsets defined by monoclonal antibodies on pokeweed mitogen stimulation in early human life. J Immunol 128:11–15

Porta FA, Maccario R, Ferrari FA, Alberini CM, Montagna D, De Amici M, Giannetti A, Ugazio AG (1985) Lymphocyte subpopulations in the neonate: high percentage of ANAE-positive cells with low avidity for sheep erythrocytes. Thymus 5:263–269

Solinger AM (1985) Immature T lymphocytes in human neonatal blood. Cell Immunol 92:115–122

Stutman O (1975) Humoral thymic factors influencing postthymic cells. Ann NY Acad Sci 249:89–104

Stutman O (1977) Two main features of T-cell development: thymus traffic and postthymic maturation. Contemp Top Immunobiol 7:1–46

Stutman O (1978) Intrathymic and extrathymic T cell maturation. Immunol Rev 42:138–184
Stutman O (1985) Ontogeny of T cells. Clinics in Immunology and Allergy 5:191–234
Toivanen P, Uksila J, Leino A, Lassila O, Hirvonen T, Ruuskanen O (1981) Development of mitogen responding T cells and Natural Killer cells in the human fetus. Immunol Rev 57:89–106
Ugazio AG, Altamura D, Giraudi D, Mingrat G, Beloni C, Burgio GR (1976) Peripheral blood T lymphocyte subpopulations in newborns. Boll Ist Sieroter Milanese 55:451–459
Vitiello A, Maccario R, Montagna D, Porta FA, Alberini CM, Mingrat G, Astaldi-Ricotti GCB, Nespoli L, Ugazio AG (1984) Lymphocyte subpopulations in the neonate: a subset of HNK-1–, OKT3–, OKT8+ lymphocytes displays natural killer activity. Cell Immunol 85:252–257
Yokoi T, Miyawaki T, Yachie A, Ohzeki S, Taniguchi N (1982) Discrepancy in expression ability of Tac antigen and Ia determinants defined by monoclonal antibodies on activated or cultured cord blood T lymphocytes. J Immunol 129:1441–1445.

Analysis of the Immunological Dysregulation Underlying Defective Interferon γ Secretion in the Human Neonate

J. L. Virelizier and N. Wakasugi

Unité d'Immunologie et d'Hématologie Pédiatriques Hôpital Necker-Enfants Malades 149, rue de Sèvres, 75743 Paris, France

Host responses to a range of microorganisms, including bacteria, mycobacteria, yeasts, parasites, and viruses, are mediocre in the human neonate. In the case of severe bacterial infections, lack of preexisting antibody (Baker and Kasper 1976), decreased complement levels (Fischer and Pearlman 1961), or deficient functions of phagocytes (Weston et al. 1977) may be responsible for the inadequate handling of the microorganism. In the case of infection with intracellular microorganisms (and particularly viruses), antibody and phagocytosis are unlikely to play a decisive role in host defense, so that a T-cell-mediated defect has to be postulated in order to explain the unusual susceptibility of neonates to such infections (Bellanti and Hurtado 1976). The so-called immaturity of cell-mediated immunity in neonates has been difficult to substantiate. Indeed, the number of monocytes, B and T lymphocytes (Campbell et al. 1974), subpopulations of T cells (Hayward and Kurnick 1981), T lymphocyte proliferation induced by mitogens (Campbell et al. 1974), or allogeneic cells (Granberg et al. 1976) are known to be normal in neonates.

Interferon (IFN) has a role in host defense against virus infections (Baron 1973) and in the regulation of immune responses (Sonnenveld 1980; Virelizier 1980). The ability of neonatal leukocytes to secrete IFN-α after nonspecific induction by Sendai (Ray 1970) or herpes simplex virus (Kohl and Harmon 1980) has been shown to be normal. In contrast, Bryson et al. (1980) have shown a profound defect of IFN-γ production after stimulation with phytohemagglutinin (PHA) of leukocytes from cord blood or peripheral blood from infants less than 1 week old. IFN-γ, apart from its antiviral effects, has intense immunoregulatory properties, and appears to behave as a macrophage activating factor (MAF), since it increases macrophage cytotoxicity (Schultz and Kleinschmidt 1983), intracellular killing, and oxidative metabolism (Nathan et al. 1983) and class II HLA-antigen expression (Virelizier et al. 1984) in human monocytes. Thus, defective production of this lymphokine may have serious immunological consequences. Indeed, the contrast between a profound defect of IFN γ secretion and normal or subnormal other immune functions in newborns is reminiscent of a similar dissociation observed in patients with selective defects of interferon secretion (Virelizier 1984) and suggests that this defect contributes to the unusual susceptibility to infections in neonates. This prompted us to investigate whether the defect is intrinsic (and thus unlikely to be amenable to therapeutic intervention) or secondary to defective regulation. Our results have shown that the IFN-γ

Immunology of the Neonate
Burgio, Hanson, Ugazio (Eds.)

defect of cord blood leukocytes (CBL) can be reversed by irradiation, by preincubation or by staphylococal enterotoxin A (SEA) stimulation, and that no active cellular suppression of IFN-γ secretion can be found in CBL. Furthermore, we have analysed the possible role of amplifiying or inhibitory mediators in the abnormal regulation leading to decreased IFN-γ secretion in CBL (Wakasugi and Virelizier 1985; Wakasugi et al. 1985).

We have shown that IFN-γ production induced by lectins such as PHA is profoundly diminished in CBL cultures. This confirms the observation of Bryson et al. (1980) that PHA-induced IFN-γ secretion is defective in CBL and human neonates. The defect in CBL can also be observed after stimulation with another lectin (ConA) or with a monoclonal antibody to T lymphocytes (OKT3), whereas similar stimulation of leukocytes from mothers at time of delivery results in normal IFN production. OKT3 stimulation induces IFN-γ in cells of the T lymphocyte lineage (Wussow et al. 1981), which suggests that T lymphocytes in CBL are defective in production of IFN-γ. TPA is known to be a costimulator of PHA in T cells, able to augment the production of interleukin 2 (IL2) and IFN-γ (Northoff et al. 1980). Indeed, we have observed that addition of TPA to PHA consistently augments the titer of IFN-γ in supernatants of adult control leukocytes. In contrast, very little, if any, increase in IFN-γ secretion was observed in CBL cultures stimulated with PHA and TPA. Thus the defective production of the lymphokine by CBL cannot be reversed by costimulation with TPA, indicating a profound unresponsiveness of IFN-γ-producing cells in CBL.

The intrinsic ability of CBL to secrete IFN-γ appears to be normal, since simple and nonaggressive manipulation of CBL can largely or completely reconstitute the ability to produce IFN-γ. Thus, irradiation of fresh CBL with as little as 500 or 1000 rads permitted a normal IFN-γ secretion in response to PHA. Furthermore, mere incubation of CBL cultures for 24 h at 37°C before addition of PHA was sufficient to reverse the defective IFN-γ production. Finally, using SEA instead of lectins led to high titers of IFN-γ in CBL cultures, comparable to those observed in cultures from controls or mothers. In every instance, the type of IFN produced was shown to be IFN-γ, with very little, if any, detectable contamination of supernatants by IFN-α or IFN-β. This indicates that the defect of IFN-γ secretion by CBL is not absolute, but rather the consequence of an abnormal regulation of T lymphocyte functions. In the case of SEA induction there is no decisive evidence that T lymphocytes are the sole producers of IFN-γ. In the case of PHA, ConA, or OKT3 induction, however, reversal of IFN-γ secretion after irradiation of incubation indicates that the intrinsic ability of cord blood T lymphocytes to produce this lymphokine is normal.

The reversibility of the IFN-γ secretion in CBL led us to look for a defective accessory function of fetal monocytes, since is has been shown that IFN-γ secretion is macrophage dependent (Epstein et al. 1971). However, the addition of adherent cells from normal donors did not reverse the IFN-γ defect of CBL. Alternatively, the defective IFN secretion could be the result of a suppressive mechanism that would be blocked by irradiation or incubation, such as an effect of suppressor T lymphocytes. Indeed, suppressor T cells able to inhibit PWM-induced maturation of B lymphocytes into antibody-producing cells have been described in CBL (Durandy et al. 1979; Hayward and Lydyard 1978; Miyawaki et al. 1981; Olding and Oldstone 1976; Rodriguez et al 1981). There is no general agreement on the nature of the T-cell subpopulation(s) involved, or on the mechanisms of CBL-induced suppression. It should be noted that active suppression of lymphokine production of normal leukocytes by CBL has not been reported. In our study, we have been unable to find evidence of active cellular suppression of the PHA-induced IFN-γ production in coculture experiments mixing adult leukocytes and CBL. We have tested the hypothesis that a putative suppressor cell in this system would be HLA-restricted in its effects. No evidence for such cells were found, since cocultures of paired mother-cord leukocytes did not result in lower IFN-γ titers than those produced by maternal leukocytes induced by PHA in the absence of cord leukocytes. Such negative results do not exclude the existence of a suppressor cell able to inhibit IFN-γ production in CBL, but clearly indicate that the effect of this putative suppressor leukocytes is not easily transferable in vitro to adult leukocyte cultures.

These observations would be compatible with a suppressive phenomenon active with CBL, but not (or much less) with adult leukocytes. However, mixtures of irradiated and nonirradiated leukocytes from individual neonates resulted in IFN-γ production comparable to that of irradiated CBL and thus similar to IFN-γ titers observed in adult leukocyte cultures. We have thus been unable to demonstrate any active cellular suppression in our cocultures, even when CBL were used in assay for a putative suppressor cell. In fact, two situations would be consistent with our results: neither defective amplification mechanisms, nor an excessive sensitivity to a suppressive mediator in CBL would be expected to result in suppression in cocultures of CBL and adult leukocytes. A systematic analysis of the mediators able to amplify immune responses and IFN-γ production (interleukin 2 and leukotrienes) or to suppress lymphokine production (prostaglandins) has been reported elsewhere (Wakasuki et al. 1985). The results are compatible with the present interpretation, and show that an exaggerated sensitivity of CBL to the suppressive effects of prostaglandins, rather than an excessive induction of a universally active suppressor phenomenon, underlies the profound defect of IFN-γ secretion in CBL.

Whatever the mechanism of the dysregulation observed in CBL, the potential relevance of the defective IFN-γ production by cord leukocytes is considerable. It is interesting to note that leukocytes from mothers bled at the time of delivery produce normal titers of IFN-γ, a finding which suggests that the profound defect of CBL is not merely a reflection of a transient hormonal change associated with delivery. Indeed, it has been shown that the IFN-γ defect persists for at least a week in neonates (Bryson et al. 1980). Thus, the defective secretion of this lymphokine is likely to have consequences in terms of immunological regulation in early infancy. At this age, the main other immunologic abnormalities are a low natural killer (NK) acitivity (Kaplan et al. 1982; Tarkkanen and Saksela 1982), and a low expression of class II HLA antigens on the membrane of T lymphocytes (Yokoi et al. 1982) and monocytes (Kelley et al. 1984). Interestingly, the low NK activity of CBL can be increased by interferon (Kaplan et al. 1982; Tarkkanen and Saksela 1982), and IFN-γ reverses the defective HLA-DR expression of fetal monocytes (Kelley et al. 1984). These observations suggest that the defective NK activity and HLA-DR expression are consequences rather than causes of the IFN-γ defect.

Finally, it should be emphasized that the defect of IFN-γ production of human neonates appears to be relatively selective even in terms of lymphokine production, since the secretion of IL2 (Hayward and Kurnick 1981) and lymphocyte-derived chemotactic factor (Keller et al. 1983) is normal in CBL, whereas the production of migration inhibition factor is reported to be subnormal at that age (Winter et al. 1983). If indeed IFN-γ behaves in vivo as a macrophage-activating factor, as is shown in vitro (Nathan et al. 1983; Schultz and Kleinschmidt 1983; Virelizier et al. 1984), the relatively selective defect of IFN-γ secretion in neonates my be envisaged as a major cause of immunological dysfunction at that age, and may contribute, through diminished activation of the reticulo-endothelial system, to the increased susceptibility of neonates to infections with intracellular microorganisms.

References

Baker CJ, Kasper DL (1976) Correlation of maternal antibody deficiency with susceptibility to neonatal group B streptococcal infection. N Engl J Med 294:753-756

Baron S (1973) The defensive and biological roles of the interferon system. Front Biol 2:267

Bellanti JA, Hurtado JA (1976) Immunology and resistance to infection. In: Remington JS, Klein JO (eds) Infectious diseases of the fetus and newborn infant. Saunders, Philadelphia, p 33

Bryson YJ, Winter HS, Gard SE, Fischer TJ, Stiem ER (1980) Deficiency of immune interferon production by leukocytes of normal newborns. Cell Immunol 55:191-200

Campbell AC, Waller C, Wood J, Aynsley-Green A, Yu V (1974) Lymphocyte subpopulations in the blood of newborn infants. Clin Exp Immunol 18:469

Durandy A, Fischer A, Griscelli C (1979) Active suppression of B lymphocyte maturation by two different newborn T lymphocyte subsets. J Immunol 123:2644-2650

Epstein LB, Cline MJ, Merigan TC (1971) The interaction of human macrophages and lymphocytes in the phytohemagglutinin-stimulated production of interferon. J Clin Invest 50:744-753

Fischer CW, Pearlman DS (1961) Complement component of paired mother-cord sera. Proc Soc Exp Biol Med 107:695

Granberg C, Manninen K, Toivanen P (1976) Cell mediated lympholysis by human neonatal lymphocytes. Clin Immunol Immunopathol 6:256

Hayward AR, Kurnick J (1981) Newborn T cell suppression: early appearance, maintenance in culture and lack of growth factor suppression. J Immunol 126:50-53

Hayward AR, Lydyard PM (1978) Suppression of B lymphocytes differentiation by newborn T lymphocytes with an Fc receptor for IgM. Clin Exp Immunol 34:374-378

Kaplan J, Shope TC, Bollinger RO, Smith J (1982) Human newborns are deficient in natural killer activity. J Clin Immunol 2:350-355

Keller MA, Kidd RM, Leake RD, Evereto SL (1983) Lymphocyte-derived chemotactic factor production by neonatal lymphocytes. Pediatr Res 17:799-802

Kelley VE, Fiers W, Strom TB (1984) Cloned human interferon γ but not interferon α or β induces expression of HLA-DR determinants by fetal monocytes and myeloid leukemic cell lines. J Immunol 132:240-245

Kohl S, Harmon MW (1980) Human neonatal leukocyte interferon production and natural killer cytotoxicity in response to herpes simplex virus. J IFN Res 3:461

Miller ME (1978) Current topics in host defenses of the newborn. In: Barness LA (ed) Advances in pediatrics, Vol 25. Year Book Medical Publishers, Chicago, p 59

Miyawaki T, Moriya N, Nagaoki T, Kubo M, Yokoi T, Taniguchi N (1981) Mode of action of humoral suppressor factor derived from pokeweed mitogen stimulated cord T cells on adult B cell differentiation. J Immunol 126:282-285

Nathan CF, Murray HW, Wiebe ME, Rubin BY (1983) Identification of interferon γ as the lymphokine that activates human macrophage oxidative metabolism and antimicrobial activity. J Exp Med 158:670-689

Northoff H, Hooks JJ, Jordan G, Oppenheim JJ (1980) Direct correlation between the production of Interleukin 2 (IL2) and immune interferon (IF) by human mononuclear cells (MNL). Behring Inst Mitt 67:90
Olding LB, Oldstone M (1976) Thymus-derived peripheral lymphocytes from human newborns inhibit division of their mother's lymphocytes. J Immunol 116:682–686
Ray CG (1970) The ontogeny of interferon production by human leucocytes. J Pediatr 76:94–98
Rodriguez MA, Bankhurst AD, Ceuppens JL, Williams RC (1981) Characterization of the suppressor cell activity in human cord blood lymphocytes. J Clin Invest 68:1577–1582
Schultz RM, Kleinschmidt WJ (1983) Functional identity between murine γ interferon and macrophage activating factor. Nature 305:239–240
Sonnenfeld G (1980) Modulation of immunity by interferon. In: Pick E (ed) Lymphokine reports, vol 1. Academic, New York, p 113
Tarkkanen J, Saksela E (1982) Umbilical cord-Blood derived suppressor cells of the human natural killer cell activity are inhibited by interferon. Scand J Immunol 15:149–157
Virelizier JL (1980) Immunological aspects of interferon secretion. In: Collier LH, Oxford J (eds) Developments in antiviral therapy. Academic, London, p 201
Virelizier JL (1984) Deficiencies of interferon production in children. Press Med 13:495–498
Virelizier JL, Perez N, Arenzana-Seisdedos F, Devos R (1984) Pure interferon gamma enhances class II HLA antigens on human monocytes cell lines. Eur J Immunol 14:106–108
Wakasugi N, Virelizier JL (1985) Defective IFN γ production in the neonate. I- Dysregulation rather than intrinsic abnormality. J Immunol 134:167–172
Wakasugi N, Virelizier JL, Arenzana-Seisdedos F, Rothhut B, Mencia-Huerta JM, Russo-Marie F, Fiers W (1985) Defective IFN γ production in the human neonate: II. Role of increased sensitivity to the suppressive effect of prostaglandin E. J Immunol 134:172–176
Weston WL, Carson BS, Barkin RM, Slater GD, Dustin RD, Hecht SK (1977) Monocyte-macrophage function in the newborn. Am J Dis Child 131:1241–1242
Winter HS, Gard SE, Fischer TJ, Bryson YJ (1983) Deficient lymphokine production of newborn lymphocytes. Pediatr Res 17:573–578
Wussow P, Platsoucas CD, Wiranowska-Stewart M, Stewart WE (1981) Human γ-interferon production by leukocytes induced with monoclonal antibodies recognizing T cells. J Immunol 127:1197–1200
Yokoi T, Miyawaki T, Yachie A, Ohzeki S, Taniguchi N (1982) Discrepancy in expression ability of Tac antigen and Ia determinants defined by monoclonal antibodies on activated or cultured cord blood T lymphocytes. J Immunol 129:1441–1445

Immunity and Infections in the Neonate

JOSEPH A. BELLANTI and BARBARA J. ZELIGS

Departments of Pediatrics and Microbiology, and The International Center for Interdisciplinary Studies of Immunology, Georgetown University School of Medicine, Washington, D.C., USA

1 Introduction

Bacterial infection of the newborn is a common problem that causes significant morbidity and mortality for the host. The overall risk of the more common types of neonatal bacterial infection - sepsis and meningitis - ranges from 3 to 6/1000 live births in the United States. Approximately 20%–50% of infected infants die of the illness, and for those who are treated successfully for meningeal infection, neurological sequelae are found in about one-half. Although a number of organisms have been linked etiologically with sepsis and meningitis neonatorum, 60% or more of these infections are primarily due to three pathogens, group B streptococcus (GBS), *E. coli,* and *L. monocytogenes.*

Epidemiological and clinical characteristics of bacterial infections in the newborn and young infant suggest that the newborn human acts as an immunologically compromised host, and that these defects are age-related. First, the overall incidence of these infections (3–6/1000 live births during the first few weeks of life) by far exceeds the risk of infection encountered during the remainder of the individual's lifespan. As the newborn matures it is rare for infection with these three pathogens to occur. Moreover, in the newborn the frequency of infections caused by the common pathogens of the older infant (i.e., *H. influenzae, S. pneumoniae* and *N. meningitidis*) is considerably lower. Second, even within the newborn period, relative rates of infection vary with the gestational and chronological age of the host. The premature infant, for example, experiences a higher attack rate for GBS and *E. coli* infection in the first days of life than the term baby. The attack rate for GBS infection in the first days of life is estimated to be 1.3/1000 to 4.1/1000 live births, a rate considerably higher than that seen in infants older than 7 days, i.e., 0.6–1.3/1000 live births.

Thus, it appears that the expressions of neonatal infections are modulated by the infant's gestational and chronological age. It is tempting to speculate that the relative resistance to infection and its beneficial consequences in the older infant are mediated, at least in part, by maturation of nonspecific mechanisms of host response. While this hypothesis has not yet been fully addressed in the human newborn host, a substantial body of data (Fischer et al. 1975; Christensen et al. 1980; Zeligs et al. 1982) is now available from experimental animal models of neonatal infection to support this view. There appears to be an age-dependent variation in inflammatory response, which correlates with the distinct age-depen-

Immunology of the Neonate
Burgio, Hanson, Ugazio (Eds.)

dent susceptibility to GBS and *L. monocytogenes*. In addition, the role of antibody in this neonatal susceptibility to infection has also been investigated. It has been shown that specific antibody can be protective (Baker 1979), and further that the administration of adult neutrophils at the time of infection offered significant protection (Laurenti et al. 1981; Christensen 1982). These data lend additional support to the hypothesis that the immaturity of the neonatal host's immune response may in part be responsible for its susceptibility to bacterial infections.

2 Components of Nonspecific Immune Response

In attempting to determine the mechanism(s) responsible for this immaturity in the newborn, several investigators have examined both nonspecific and specific immunologic parameters. Although it is unclear which of these systems is the more important in preventing infection, nonspecific immune responses appear to play a more critical role in preventing initial infection. The events comprising the nonspecific responses involve two major cell types: (a) the polymorphonuclear leukocyte (PMN), and (b) the mononuclear phagocyte, which consists of the circulating monocyte as well as tissue macrophages. The responses of these cells during inflammation consist of a series of chronologic events beginning with their initial influx into the site of inflammation, i.e., chemotaxis. This influx occurs in response to a series of factors referred to as chemoattractants, which include both cellular factors and complement components which are generated in response to microorganisms. The second event consists of adherence and ingestion (phagocytosis) of the invading organisms which are enhanced by opsonizing factors, e.g., complement and antibody. The final phase of the inflammatory response culminates in the killing and degradation of the organism. These events are accompanied by a series of complex metabolic events, which include the activation of glycolytic pathways, the generation of activated forms of oxygen, and increased utilization of oxygen, events collectively referred to as the "respiratory burst." If complete elimination of the microorganism does not occur initially, the bone marrow reserves are stimulated to increase proliferation and differentiation of precursor cells from pluripotential stem cells.

2.1 Chemotactic Activity

The functional capabilities of these phagocytic cells have been examined in the young animal to further elucidate mechanisms of age-dependent susceptibility to infection. Several laboratories have reported that the chemotactic responsiveness of peripheral blood phagocytic cells in the newborn infant appears to be less than that of the adult. Klein et al. (1977) showed by chronological assessment that there was a definite maturation in chemotactic function of both PMNs and monocytes from humans. Our studies of cord blood PMNs (Boner et al. 1982) have shown defects in chemotaxis correlated with PMN differentiation, i.e., 30% of the neutrophils in cord blood were of the band form, as against 9% in adults.

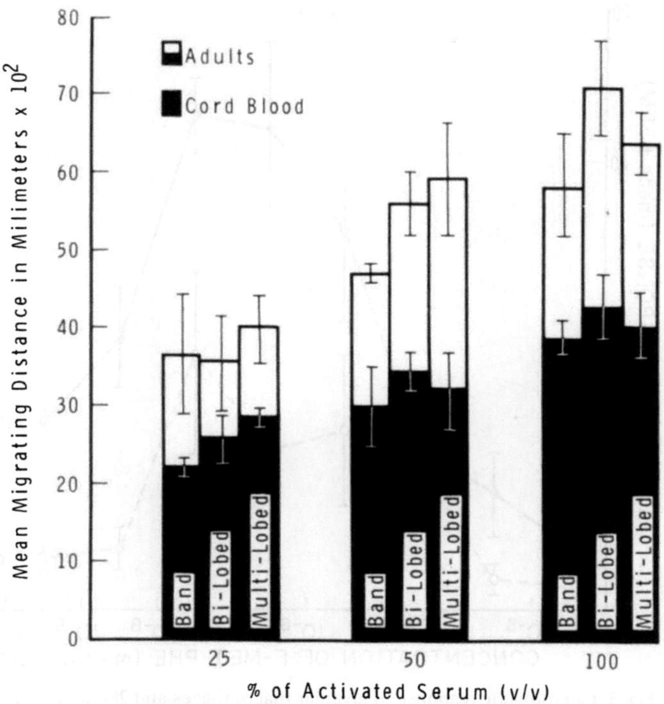

Fig. 1. Mean migrating distance of various differentiational stages of neutrophils from human adults and cord blood in response to various concentrations of activated serum. (From Boner et al. 1982)

These less differentiated PMNs had a decreased chemotactic response, migrating only small distances, as well as requiring higher concentrations of chemoattractant to respond (Fig. 1). In addition to these decreased chemotactic responses a decreased capacity of newborn sera to generate chemotactic factors was found (50% of adult levels). Further, it has been reported by Kretschmer et al. (1976) that lymphocytes from newborns have a decreased production of lymphocyte-derived chemotactic factor. It therefore appears that both the chemotactic responsiveness and some of the systems involved in the generation of chemoattractants are immature in the young host.

The cellular mechanism(s) responsible for this immaturity in chemotactic function are as yet unclear, but two responsibilities have been suggested. Firstly, the components involved in producing and sustaining the metabolic events, i.e., the respiratory burst necessary for chemotaxis, may be immature (Mills et al. 1979) as described below. Alternatively, the cell surface receptors which interact with the chemoattractants may be decreased in number. In support of the latter possibility, Klempner and Gallin (1978) have reported subpopulations of PMNs in peripheral blood which can be distinguished from one another by the presence of Fc receptors as well as by specific functional differences. These workers have reported a correlation between the presence of Fc receptors and certain functional capacities, e.g., adherence, chemotaxis, and microbicidal activity, and have demonstrated a decreased chemotactic function by cells showing fewer Fc

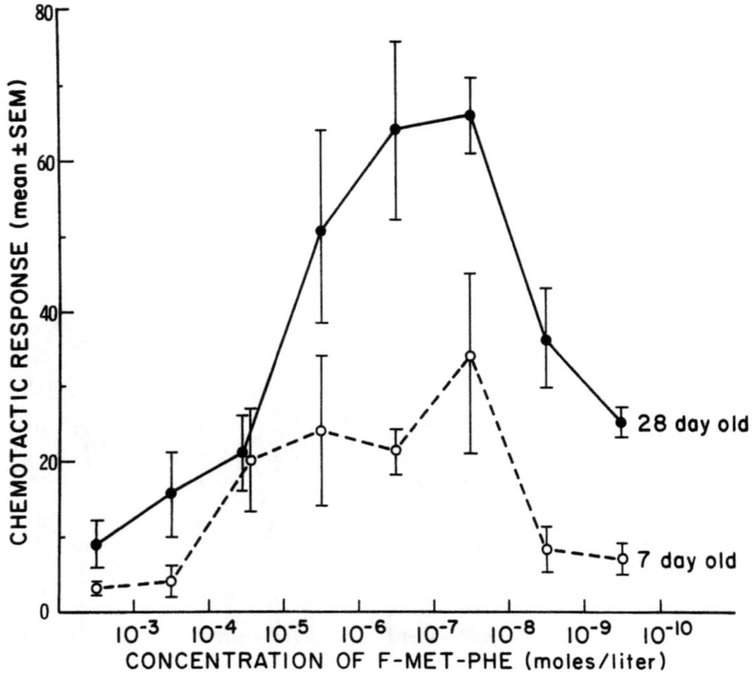

Fig. 2. Chemotactic response of alveolar macrophages and 28-day-old rabbits to various concentrations of F-MET-PHE. (From Zeligs et al. 1984)

receptors. In contrast, Strauss and Snijder (1984) have reported that both the binding and affinity of chemotactic peptide is similar in neutrophils from neonates and adults. Studies from our laboratory have shown that there is also a maturational increase in the chemotactic response and responsiveness of the alveolar macrophage (Fig. 2). This may be partially related to the suppressive effect of the large amount of surfactant released in the alveoli with birth (Zeligs et al. 1984). Thus, it appears that the first phase of inflammation, chemotaxis, is immature in the young animal.

2.2 Phagocytic Activity

Of equal importance for understanding of age-dependent susceptibility to infection is phagocytosis. There is now general agreement that phagocytic function of PMNs and monocytes in the newborn is equal to that of the adult (Mills et al. 1979). In our laboratory we have demonstrated that phagocytic function of alveolar macrophages does not change after birth (Zeligs et al. 1977), an observation which correlates with phagocytic studies of peritoneal and tissue macrophages (Reade and Casley-Smith 1965).

2.3 Opsonic Activity

Several investigators have shown opsonic deficiency in neonatal sera (Miller 1969). Maternally derived immunoglobulin G provides some opsonic activity; however, fibronectin and complement deficiency, particularly C3 and alternative pathway components, contribute to the inadequate opsonic activity of the newborn (Norman et al. 1975; Adamkin et al. 1978; Gerdes et al. 1983). Wright et al. (1975) have demonstrated that PMNs from newborns stressed from either infectious or metabolic causes may have depressed phagocytic function even when tested in the presence of adult sera. Thus, humoral factors appear to be the major critical determinant of decreased phagocytosis except under certain conditions, such as stress, when the cell itself may be dysfunctional.

2.4 Microbicidal Activity

The final event of the nonspecific response is microbicidal activity of the phagocyte. Numerous reports have appeared in the literature suggesting that normal term infants have intact intracellular killing mechanisms, i.e., by both PMNs and monocytes. Mills et al. (1979) have demonstrated some important differences in the bactericidal activity of PMNs between healthy neonates and adults. At a low bacteria-to-PMN ratio (1:1) microbicidal activity was similar; however, at higher bacteria-to-PMN ratios (100:1) there was significantly decreased killing for *E. coli* and *S. aureus* by newborn leukocytes. Reports from our laboratory (Zeligs et al. 1977a) have described decreased microbicidal activity of alveolar macrophages in the newborn rabbits, which increases with postnatal development (Fig. 3). Collectively, these results correlate with in vivo studies of peritoneal and tissue macrophages which show postnatal maturation of bactericidal mechanisms.

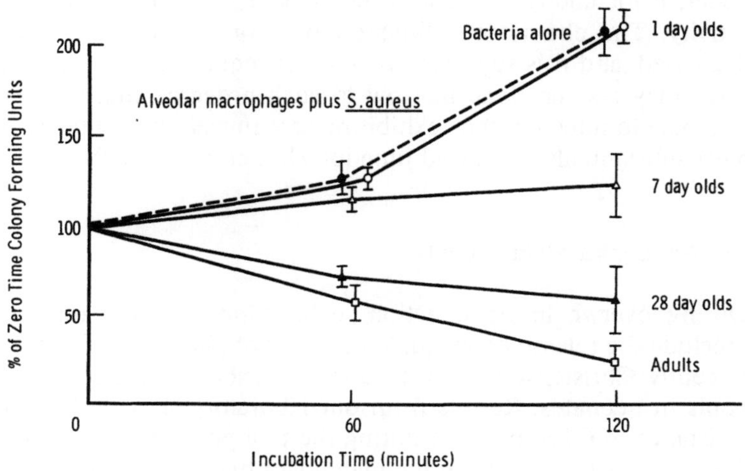

Fig. 3. Bactericidal activity of alveolar macrophages from 1, 7, and 28-day-old rabbits. (From Zeligs et al. 1977a)

Moreover, postnatal development of bactericidal mechanisms in macrophages is also suggested by the reported capacity of adoptively transferred adult peritoneal macrophages to protect newborn animals against otherwise lethal infections (Blaese 1975). It therefore appears that there may be an immaturity of the microbicidal mechanisms of the phagocytic cells of the neonate. The precise mechanism responsible for this immaturity in function is as yet unclear, but studies of oxidative metabolic pathways of the neonatal phagocytes shed light on these decreased bactericidal responses.

2.5 Metabolic Activity

Studies from our laboratory have shown maturational increases in activity and stability of enzymes of the oxidative burst in both the neutrophil (Bellanti et al. 1970) and the alveolar macrophage (Nerurkar et al. 1977) which correlate with maturation of bactericidal activity. Recently, Ambruso et al. 1984) have shown decreases in lactoferrin and lysozyme, two specific granule components involved in bactericidal activity, and hydroxyl radical formation, a potent antimicrobial substance in cord blood neutrophils. These results suggest a biochemical basis for the decreased bactericidal activity of the neonatal cells. Further, recent studies have reported a decreased viability in the neutrophils from human infants (Strauss and Snijder 1981). This has been correlated with a decrease of glutathione peroxidase and catalase in these cells, enzymes responsible for detoxifying H_2O_2, which if not removed, may cause auto-oxidation resulting in cell damages and death. The addition of superoxide dismutase and catalase increased their viability, further supporting this theory.

Although the newborn phagocyte is metabolically active, it may nonetheless be near the limits of its capability secondary to birth. The studies of Shigeoka et al. (1980) have demonstrated that PMNs with additional stress or infection have deficient metabolic response to particulate material and impaired microbicidal activity. The chemotactic function of neonatal cells has been shown to be decreased, and it is suggested that stress requiring rapid recruitment of phagocytes may tax the bone marrow to such a degree that the cells generated in response to infection may exhibit mature nuclear morphology but, in fact, may have substantially decreased function (Boner et al. 1982).

2.6 Neutrophil Storage Pools

During overwhelming infection, exhaustion of bone marrow reserves may preclude the release of adequate numbers of phagocytes from the bone marrow. Recently, Christensen and Rothstein (1980) showed decreased neutrophil storage pools in neonates. Results from our laboratory show a day-by-day increase in resistance to GBS infection during the first postnatal weeks (Zeligs et al. 1982). This appears to result from the inability of the younger animals to mobilize cells from their neutrophil storage pools as reflected in the severe neutropenia sustained by the infected newborns compared to 7-day-old animals infected with a similar dosage of GBS. Several factors which we have found to change with animal maturation may play a role in age-dependent susceptibility.

These include the ability to perpetuate the neutrophil storage pools and the size of these pools; the delay in the mobilization of peripheral blood leukocytes to site of infection; the decreased chemotactic response of cord blood and newborn cells compared to adults; and the lower capacity of newborn serum to generate chemotactic factors.

3 Factors Contributing to Neonatal Phagocytic Cell Dysfunction

Our team has carried out a series of studies to investigate parameters which may play a role in the susceptibility of the newborn to infection. These studies have concentrated on several factors which may directly result in the decreased function of the phagocytic cells, either by interfering with the function of the phagocytic cells, either by interfering with the function of these cells directly or by inducing extracellular conditions which result in deficiencies in the total cell population, or both.

3.1 Surfactant

The initial studies involved the investigation of factors which appear to directly induce a dysfunction in the phagocytic cells. The first of these was a continuation of studies which focused on the alveolar macrophage. Our previous studies of the lung (Zeligs et al. 1977a, 1977b, 1984; Bellanti et al. 1985), a major target of infection in the infant, had suggested that the initial response of the alveolar macrophage (AM), the predominant phagocytic cell of this site, was impaired. This appeared to be due to both an intrinsic immaturity in the AM as well as the influence of extracellular factors present in the lung in the early postnatal period. These factors appeared to be associated with the large amount of surfactant-related material (SRM), surface active phospholipids which are important in maintaining functional alveoli. Large quantities of SRM are released at birth and are taken up by AM at this time. The results of these studies showed that the chemotactic and candidacidal response of AM increases with postnatal maturation, i.e., responses of AM from 1-day-old rabbits, $<$7-day rabbits, $<$28-day rabbits. In addition, the chemotactic and candidacidal responses of AM from 7- and 28-day-old rabbits were decreased significantly if cells were pretreated with liposomes of either surfactant or its major phospholipid, dipalmitoyl lecithin. These results suggested that the decreased function of the AM in the newborn may be related to both an immaturity in the cell and to effects of the large amount of surfactant-related material present at this time (Zeligs et al. 1984).

3.2 Arachidonic Acid and Oxidative Burst Metabolites

The initial studies compared the production of the arachidonate cyclooxygenase metabolites, thromboxane (TXB2) and prostaglandin E_2 (PGE_2), from the neutrophils of adult human males and females (Mallery, S., Zeligs, B.J., Ramwell, P., Bellanti, J.A., 1985, unpublished work). The rationale of these studies was based on the known increased susceptibility to infection which has been reported in males. These observations may also be of relevance to the higher incidence of infection in the male neonatal infant. Assays were done to study the reactive oxygen species (ROS), superoxide, a key metabolite in the oxidative burst. The results show an increased production of TxB_2 and PGE_2 with a decreased production of O_2 in men's neutrophils compared with levels from women's cells. The lower levels of arachidonate metabolites in women appear to be related to the increased oxidative burst, since the TxB_2 levels increased with a ROS scavenger. Further, there was a variation in the levels of metabolites from these two systems in women with their menstrual cycle, which was not seen in men. These variations suggest that at times when estrodial levels are higher the oxidative burst is enhanced with a concomitant decrease in TxB_2 production, and vice versa. In addition, PGE_2 has been reported to decrease some phagocytic cell functions.

These results suggest an important regulatory interaction between the arachidonate and oxidative burst metabolism, which may be partially modulated by hormones. It therefore seems possible that preferential activation of one or the other of these systems may result in either an appropriate or an inappropriate inflammatory response to an infecting organism.

3.3 Protein-Calorie Malnutrition

Studies were performed to evaluate the effects of protein-calorie malnutrition on phagocytic cell development and function (Zeligs, B.J., Bellanti, J.A, 1985, unpublished work). These studies are based on the known increased susceptibility to infections of infants who have had poor antenatal and/or postnatal nutrition, i.e., premature and small for gestational age infants; drug-addicted, alcoholic, or Third World country mothers. These studies utilized rats which had been placed on protein-deficient diets either within the 1st week or after 1 month of life. The quantity and quality of the neutrophil storage pools, i.e., bone marrow, spleen, and peripheral blood, were examined and compared with values of normal animals. In addition, the susceptibility of these animals to two common pathogens of the neonatal period, group B streptococci and *E. coli,* were evaluated. The results of these studies showed that when litters of rats received protein-deficient diets within the 1st week of life the neutrophil pools were severely decreased, a finding which correlated with an increased susceptibility to infection. Conversely, when the protein-deficient diet was administered to older animals, the effects were minimal. These results provide evidence that protein deprivation has an effect on phagocytic cell function and that these effects may be more profound during early stages of postnatal maturation.

3.4 Microbial Toxins

Products released from microorganisms also appear to play a role in the susceptibility of the newborn to bacterial infections. Recently, a number of studies have been directed toward the role of the specific organism and its products in mediating sepsis. Specific toxins from group B streptococci have been reported to induce many clinical manifestations of infections (Rojas et al. 1981; Warejeka et al. 1985). The pathogenicity of this organism appears to be related to its inability to activate the alternative pathway of complement significantly. This pathway plays a major role in the clearance of this organism, and its components are reduced in this age group.

4 Summary

Current data suggest that the susceptibility of the newborn to bacterial infections is related to a wide variety of factors, which result in an inability of the newborn to mobilize and maintain the numbers of inflammatory cells necessary for clearance of the organisms. This appears to be due to immaturity of both cellular and humoral components as well as to modulation by extracellular products. These include bacterial toxins, sex hormones, metabolites released by the cells themselves, and nutritional factors, e.g. dietary protein. This hypothesis is further substantiated by recent therapeutic studies which have shown increased survival in neonatal sepsis after transfusions with adult leukocytes (Laurenti et al. 1981; Christensen et al. 1982), specific immune serum (Fischer et al. 1982), and non-steroidal anti-inflammatory drugs (Short et al. 1982). These observations should provide new insight into the prevention and therapy of infections in the human neonatal infant.

References

Adamkin D, Stitzel A, Urmson J, Farnett ML, Post E, Spitzer R (1978) Activity of the alternative pathway of complement in the newborn infant. J Pediatr 93:604–608

Ambruso DR, Bentwood B, Henson PM, Johnson RB (1984) Oxidative metabolism of cord blood neutrophils: Relationship to content and degranulation of cytoplasmic granules. Pediatr Res 18:1148–1153

Baker CJ (1979) Group B Streptococcal Infections in Neonates. Pediatrics in Review 1:5–15

Bellanti JA, Cantz BE, Schlegel RJ (1970) Accelerated decay of glucose-6-phosphate dehydrogenase activity in chronic granulomatous disease. Pediatr Res 4:405

Bellanti JA, Zeligs BJ, Sordelli DO (1985) Pulmonary immunodeficiency. Int Arch Allergy Appl Immunol 76: [Suppl 1]:101–107

Blaese MR (1975) Macrophages and the development of immunocompetence. In: Bellanti JA and Dayton DH (ed) The phagocytic cell in host resistance. Raven, New York, p 309–315

Boner A, Zeligs BJ, Bellanti JA (1982) Chemotactic responses of various differentiational stages of neutrophils from human cord and adult blood. Infect Immun 35:921–928

Christensen RD, Bradley PP, Priebat DA, Anstall H, Rothstein G (1982) Granulocyte transfusions in septic canine neonates. Pediatr Res 16:57–59

Christensen RD, Rothstein G (1980) Efficiency of neutrophil migration in the neonate. Pediatr Res 14:1147–1149

Christensen RD, Shigeoka AO, Hill HR, Rothstein G (1980) Circulating and storage neutrophil changes in experimental type II Group B Steptococcal sepsis. Pediatr Res 14:806–808

Fisher GW, Podgore JK, Bass JW, Kelly JL, Kobayasnki GY (1975) Enhanced host defense mechanisms with Levamisole in suckling rats. J Infect Dis 132:578–581

Fischer GW, Wilson SM, Hunter KW (1982) Functional characteristics of a modified immunoglobulin preparation for intravenous administration: Summary of studies of opsonic and protective activity against group B streptococcus. J Clin Immunol 2:31S–35S

Gerdes JS, Yoder MC, Douglas SD, Poles RA (1983) Decreased fibronectin in neonatal sepsis. Pediatrics 72:877–881

Klein RB, Fischer TJ, Gard SE, Biberstein B, Rick KC, Stiehm ER (1977) Decreased mononuclear and polymorphonuclear chemotaxis in human newborn, infants and young children. Pediatrics 60:467–472

Klempner MS, Gallin JL (1978) Separation and functional characterization of human neutrophil subpopulations. Blood 51:659–669

Kretschmer RR, Stewardson PB, Papierniak CK, Gotoff SP (1976) Chemotactic and bactericidal capacieties of human newborn monocytes. J Immunol 119:1303–1307

Laurenti F, Ferro R, Isacchi G, Panero A, Savignoni PG, Malognino F, Palermo D, Mandelli F, Bucci G (1981) Polymorphonuclear leukocyte transfusion for the treatment of sepsis in newborn infant. J Pediatr 98:118–123

Miller ME (1969) Phagocytosis in the newborn infant: humoral and cellular factors. J Pediatr 74:255–259

Nerurkar LS, Zeligs BJ, Bellanti JA (1977) Maturation of the rabbit alveolar macrophage during animal development: II. Biochemical and enzymatic studies. Pediatr Res 11:1202–1209

Norman ME, Gall EP, Taylor A, Laster L, Nilsson UR (1975) Serum complement profiles in infants and children. J Pediatr 87:912–916

Reade PC, Casley-Smith JR (1965) The functional development of the reticulo-endothelial system: II. The histology of blood clearance by the fined macrophages of foetal rats. Immunology 9:61–66

Rojas J, Green RS, Hellerqvist RO, Brigham KL, Stahlman MT (1981) Studies on group B-hemolytic streptococcus II: Effects on pulmonary hemodynamics and vascular permeability in unanesthetized shepp. Pediatr Res 15:899–904

Shigeoka AO, Charette RP, Wyman ML, Hill HR (1980) Defective oxidative metabolic responses of neutrophils (PMNs) from stressed neonates. Annual Meeting of APS and SPR (San Antonio) April 29–May 2, Abstract # 934

Short BL, Miller MK, Fletcher JR (1982) Improved survival in the suckling rat model of group B streptococcal sepsis after treatment with non-steroidal anti-inflammatory drugs. Pediatrics 70:343–347

Strauss RG, Snijder EL (1981) Neutrophils from human infants exhibit decreased viability. Pediatr Res 15:794–797

Strauss RG, Snijder EL (1984) Chemotactic peptide binding by intact neutrophils from human neonates. Pediatr Res 18:63–66

Warejcka DJ, Goodrum KJ, Spitznagel JK (1985) Toxicity of group B *Streptococcus agalactiae* in adult rats. Infect Immun 48:560–564

Wright WC, Ank BJ, Herbert J, Stiehm ER (1975) Decreased bactericidal activity of leukocytes of stressed newborn infants. Pediatrics 56:579–584

Zeligs BJ, Nerurkar LS, Bellanti JA (1977a) Maturation of the rabbit alveolar macrophage during animal development: III Phagocytic and bactericidal functions. Pediat Res 11:1208–1215

Zeligs BJ, Nerurkar LS, Bellanti JA, Zeligs JD (1977b) Maturation of the rabbit alveolar macrophage during animal development: I. Perinatal influx in alveoli and ultrastructural differentiation. Pediatr Res 11:197–208

Zeligs BJ, Armstrong CD, Walser JB, Bellanti JA (1982) Age dependent susceptibility of neonatal rats to group B streptococcal type III infection: Correlation of severity of infection and response of myeloid pools. Infect Immun 37:255–263

Zeligs BJ, Nerurkar LS, Bellanti JA (1984) Chemotactic and candidacidal responses of rabbit alveolar macrophages during postnatal development and the modulating rules of surfactant in their response. Infect Immun 44:379–385

Intravenous Gammaglobulin Replacement for Prophylaxis of Infection in Preterm Neonates

GAETANO CHIRICO[1], GIORGIO RONDINI[1], ALESSANDRO PLEBANI[1], ALBERTO CHIARA[1], and ALBERTO G UGAZIO[2].

Division of Neonatal Intensive Care, IRCCS Policlinico San Matteo, Pavia, and Department of Pediatrics, University of Pavia[1], and University of Brescia[2], Italy

1 Introduction

Despite the improvements made in medical care in recent years, infection remains a major problem in newborn infants, because rapid diagnosis is difficult and therapy is not always effective (Eisenfeld et al. 1983). Preterm infants with gestational age ≤ 34 weeks and birth weight ≤ 1500 g are particularly at risk: the incidence of systemic infection in these infants, in fact, is about 32%, with a mortality rate of 11% (Usher 1981). The high susceptibility to infections is mainly due to the neonatal impairment of host defenses; both the aspecific and antigen-specific components of the immune system are deficient in the preterm neonate. The antibody deficiency is one of the most important contributory factors in the neonate's high susceptibility to infections: levels of IgG, all acquired via transplacental passage from the mother, are, in fact, definitely low and often similar to those found in patients with congenital agammaglobulinemia. This is because the transplacental passage occurs almost exclusively during the last 6 weeks of gestation (Hobbs and Yeung 1968; Pilgrim et al. 1975).

Other factors involved in the antibody deficiency of prematurity are deficient antibody response (particularly to lipopolysaccharide antigen and to group B streptococci), with delayed switch from IgM to other immunoglobulin isotypes, due to immaturity of B and T lymphocytes and of monocytes-macrophages (Hayward and Lydyard 1979; Cooper et al. 1982); lack of IgA and IgM, which do not cross the placenta; possible deficiency of pathogen-specific IgG, such as antibody to group B streptococci (Christensen et al. 1984; Boyer et al. 1984; Fischer et al. 1984); possible deficiency of IgG subclasses (in newborn from mother lacking one or more IgG subclasses) (Oxelius 1984). The interplay of these factors with the neonatal deficiency of complement activity (Notarangelo et al. 1984) and of neutrophil activity (Sacchi et al. 1982; Chirico et al. 1983a, b, 1985) results in an increased susceptibility to systemic infections owing to encapsulated pathogens such as group B streptococci, staphylococci, and *Klebsiella,* which require opsonization for efficient phagocytosis and killing.

The observation that the neonatal opsonic deficiency is partly correctable with gammaglobulin (Forman and Stiehm 1969) led Amer et al. (1963) and Steen (1960) to study the intramuscular use of standard human immune serum globulin in the prophylaxis of infections in premature infants. The results of these studies, however, failed to show a beneficial effect and indicated that intramus-

Immunology of the Neonate
Burgio, Hanson, Ugazio (Eds.)

cular gammaglobulin has no proven value in the routine management of low-birth-weight infants (Stiehm 1979). This lack of effect is probably due to the disadvantages of intramuscular gammaglobulin: uneven absorption, slow availability, tissue damage, and pain incurred by treatment, which makes it difficult to administer suitable amounts of IgG (Dwyer 1984).

Intramuscular preparations of gammaglobulin cannot be administered intravenously because they contain aggregates of IgG, which activate complement and may cause dangerous anaphylactoid reactions. The first attempts to achieve intravenous tolerance of gammaglobulin with enzymatic treatment resulted in alteration of the Fc fragment of the IgG, with loss of some of antibody biologic activities. During the last few years preparations of gammaglobulin have been obtained by removal of aggregates (Table 1) which can be safely administered intravenously, contain intact antibody molecules, and maintain normal biologic activities of the Fc fragment: complement activation (Romer et al. 1982b), binding to cell-surface receptors, catabolization (Morell et al. 1980), opsonic activity (Hill et al. 1984; van Furth et al. 1984). Furthermore, these preparations maintain IgG subclass distribution similar to that found in normal serum (Römer et al. 1982a, b) and contain antibodies to most of the pathogens commonly found in neonatal infections (Von Muralt and Sidiropoulos 1981). The aim of the present controlled, randomized study was to evaluate the safety and effectiveness of intravenous intact gammaglobulin substitution therapy for prophylaxis of infection in preterm neonates.

2 Patients and Methods

We studied preterm infants with gestational age $\leq$ 34 weeks and birth weight $\leq$ 1500 g. We excluded from the study infants with severe malformation, infants who died within 24 h of birth, and infants who were already infected on admission to the neonatal intensive care unit. Within 24 h of birth, infants were allocated to either treatment (intravenous gammaglobulin) or control group by the sealed envelope method. The infants of the treatment group received intravenous intact gammaglobulin (obtained at pH 4 with traces of pepsin, Sandoglobulin) at a weekly dose of 0.5 g/kg for 1 month. We used this dosage schedule on the basis of a preliminary study on the pharmacokinetics of intravenous IgG in preterm neonates (Chirico et al. 1983a, b) which showed that the serum IgG level increase after administration of 0.5 g/kg intravenous gammaglobulin is about 600–700 mg/dl and that plasma half-life of IgG is shorter in newborns (about 1 week) than in adults.

Table 1. Procedures used to achieve intravenous tolerance of gammaglobulin. (Romer et al. 1982a)

1. Enzymatic treatment: plasmin or pepsin degradation
2. Chemical modification: β-propiolactone treatment, sulfonation, reduction and alkylation
3. Removal of aggregates: pH 4 treatment with traces of pepsin, treatment with PEG, cold ethanol fractionation, adsorption by DEAE-Sephadex and protection by albumin

Serum IgG levels and type-specific antibody to group B streptococci, *E. coli* and cytomegalovirus (CMV) were evaluated (by immunodiffusion and ELISA, respectively) before and at 2, 24, 96, 168 h after administration of gammaglobulin. Infants receiving gammaglobulin were closely monitored for anaphylactoid or inflammatory reactions. Serum transaminases were evaluated in all infants at 4 months of age.

Infants were nursed in incubators; assisted ventilation was given to infants with severe RDS (respiratory distress syndrome) with pressure-limited, time-cycled, continous-flow ventilators (Bourns BP 200); transcutaneous PO_2, heart and respiratory rate, blood pressure, and cutaneous temperature were continuously monitored in all infants during the intensive care period; an umbilical-artery catheter was inserted in infants requiring a prolonged inspired oxygen concentration of greater than 60%; serial ultrasound scans of the brain were performed to detect CNS bleeding. Methods of feeding included orogastric intermittent or continuous gavage; milk formula was used when mother's milk was not available. Partial or total parenteral alimentation was used in infants who did not tolerate enteral feeding. Gestational age was determined by maternal dates and Dubowitz (1971) assessment.

Diagnosis of infection was made on the basis of microbiologic results, clinical manifestation, and nonmicrobiologic laboratory indicators (Marks and Welch 1981): total and differential white blood cell count, erythrocyte sedimentation rate, test of hemostatic mechanisms (fibrinogen, platelet count, prothrombin time, fibrin degradation products) (Corrigan 1981). Criteria for diagnosis of sepsis, menigitis, arthritis, pneumonia, urinary tract infection, and surface infection included both a positive culture of blood, cerebrospinal fluid, tracheal aspirate, urine or pus, and the presence of clinical and nonmicrobiologic laboratory features (Hemming et al. 1976). Necrotizing enterocolitis was diagnosed when typical clinical and radiologic symptoms were present (Santulli et al. 1975). Bacterial growth in cultures from infants without clinical or laboratory symptoms of infection were considered contaminants. Wide-spectrum antibiotic therapy (ampicillin and gentamicin) was started as soon as an infection was suspected, after a full septic workup, and modified with reference to the microbiologic results if necessary. A postmortem examination was performed in all infants who died.

Statistical analysis was carried out using Pearson's chisquare test or Fisher's exact test to determine the significance of differences in incidence between groups, and Student's unpaired *t*-test for comparison of means. Informed consent was obtained from at least one parent.

3 Results

Eighty-six infants admitted to the neonatal intensive care unit of Pavia between 15 January 1983 and 15 July 1985 met the criteria for enrollment in the protocol; data of three infants of the control group who died within 24 h of birth were excluded from the analysis.

Table 2. Serum levels of total IgG, anti-streptococci IgG, anti-*E.coli* IgG and anti-CMV IgG before and after administration of the first dose of intravenous gammaglobulin (0.5 g/kg) in the treated infants. Results are expressed as means ± SD

	Total IgG (mg/dl)	Anti-Strept. (AU/ml)[a]	Anti-*E.coli* (AU/ml)[a]	Anti-CMV (AU/ml)[a]
Before	616 ± 215	212 ± 246	38 ± 39	211 ± 219
2 h	1342 ± 344	1509 ± 1145	151 ± 63	608 ± 460
24 h	981 ± 187	713 ± 665	75 ± 57	458 ± 340
96 h	851 ± 175	510 ± 394	54 ± 36	411 ± 216
168 h	822 ± 190	381 ± 351	47 ± 41	328 ± 199

[a] *AU*, arbitrary unit ; 1 AU = 1/1000 of the antibody titre observed in a serum pool from normal adult controls

Mean serum IgG and type-specific antibody to group B streptococci, *E. coli* and CMV levels before and after gammaglobulin administration are shown in Table 2. Mean serum IgG levels at 2 h after gammaglobulin administration were similar to those found in normal full-term newborns. Decrement of the serum levels of total IgG, anti-streptococci, anti-*E. coli* and anti-CMV IgG was biphasic: the rapid reduction observed during the 1st day after administration related to distribution of IgG between intra- and extravascular compartments, while the slower decrement was mainly due to IgG catabolic rate. The half-life of intravascular survival, calculated from the final slope of the plasma curves, was 245 h for total IgG, 92 h for anti-streptococcal IgG, 74 h for anti-*E. coli* IgG, and 108 h for anti-CMV IgG.

No side effects attributable to gammaglobulin therapy were observed in the treated infants. No difference of serum transaminase values was recorded between treated and control infants.

Main clinical features, incidence of infection, and mortality rate of treated and control infants are summarized in Tables 3 and 4. No significant difference between the treated and control infants was noted for the factors listed in Table 3. A statistically significant reduction in the incidence of infections and of sepsis was observed in the infants who received gammaglobulin prophylaxis as compared with the controls (Table 4). Moreover, the treated infants had a reduced incidence of necrotizing enterocolitis (NEC), required fewer days of

Table 3. Clinical data of the treated and control infants

	Treated	Control	*p*
Patients *(n)*	43	40	
Mean birth weight (g)	1104	1157	NS
Mean gestational AGE (weeks)	29.7	29.8	NS
Male/female	23/20	18/22	NS
APGAR < 8 at 5 min	32(74%)	24(60%)	NS
RDS	19(44%)	16(40%)	NS

NS, nonsignificant

Table 4. Infections, clinical data, and mortality rate in the treated and control infants

	Treated	Control	*p*
Patients *(n)*	43	40	
Infections	22 (51%)	31 (77%)	<0.02
Sepsis	2 (5%)	8 (20%)	<0.05
NEC	1 (2%)	4 (10%)	NS
Intubation (days)	15.2	20.3	NS
Antibiotics (days)	24.2	30.6	NS
Age at discharge (days)	84.1	91.2	NS
Death	7 (16%)	13 (32%)	NS

NEC, necrotizing enterocolitis; NS, nonsignificant

intubation and of antibiotic therapy, were discharged earlier, and had a lower mortality rate (Table 4). These differences, however, were not significant.

A significant difference was noted in mortality rate when the main cause of death was considered (Table 5): In fact there were fewer deaths from infections among treated infants than among controls. Clinical features of infants with sepsis are reported in Table 6: While the course of sepsis was uncomplicated in the two infants who received gammaglobulin prophylaxis, severe complication and/or death was observed in most of the control infants with sepsis.

Table 5. Main cause of death in the treated and control infants

	Treated	Control	*p*
CNS bleeding, RDS	6 (14%)	7 (17%)	NS
Infection (sepsis, NEC, pneumonia)	1 (2%)	6 (15%)	<0.04

RDS, respiratory distress syndrome; NEC, necrotizing enterocolitis; NS, nonsignificant

Table 6. Clinical features in infants with sepsis

No	BW (g)	GA (weeks)	Pathogen	Age (days)	Complications
1[a]	980	28	*Staphylococcus*	11	
2[a]	1250	34	*E.coli*	24	
3	790	25	*Pseudomonas*	32	Meningitis, died
4	840	27	*Streptococcus*	100	
5	1170	32	*Staphylococcus*	45	NEC
6	1380	30	*Staphylococcus*	35	
7	1400	32	*Staphylococcus*	50	
8	1410	34	*Serratia*	12	Died
9	1500	34	*Staphylococcus*	7	NEC
10	1500	33	*Serratia*	11	

NEC, necrotizing enterocolitis; BW, body weight; GA, gestational age

[a] Infants receiving i.v. IgG prophylaxis

4 Discussion

A significant reduction in the incidence and severity of sepsis and in the mortality rate was observed in preterm, very low birth weight infants who received intravenous gammaglobulin prophylaxis. These results are apparently at variance with those reported by Amer et al. (1963) and by Steen (1960), who failed to show a preventive effect with the use of intramuscular gammaglobulin in premature infants. However, because of the intramuscular route used, the dose (3 ml/kg 16.5% solution monthly and 0.5 ml/kg every 2 weeks) was unlikely to have significantly enhanced the low IgG levels of preterm infants. Our study has confirmed that IgG levels in preterm infants are definitely low and similar to those found in patients with congenital agammaglobulinemia. Indeed, the immunodeficiency of prematurity cannot be considered physiological: in spite of the most advanced techniques of intensive care, about one-third of extremely premature infants suffer from severe infections, and 11% of them die (Usher 1981). Therefore, our study was aimed at evaluating the effect of a substitution treatment on the incidence of infections. We observed that IgG levels similar to those of term neonates could be obtained only with high-dose gammaglobulin therapy (0.5 g/kg). Whether lower doses can be sufficiently protective remains to be established.

The finding that exogenous intact gammaglobulins are rapidly catabolized by the neonate with a half-life of 10 days, as opposed to an average half-life of 3 weeks in adults, prompted us to adopt a weekly schedule of administration. The short intravascular half-life of intravenous gammaglobulin in neonates may be partly due to increased catabolic rate resulting from the relatively high serum level of IgG obtained after gammaglobulin administration, to consumption of antibody caused by the frequent neonatal infections, or to loss of proteins in infants with intestinal or renal disease.

We did not observe short-term side effects after gammaglobulin administration. As for long-term side effects, von Muralt and Sidiropoulos (1981) found no significant difference in serum IgG, IgA, and IgM levels, in response to vaccination or to skin tests for cutaneous delayed hypersensitivity, or in anti-IgG antibody level between a group of infants at 2–3 years of age who had received high doses of intravenous gammaglobulin during the neonatal period for treatment of sepsis and control infants. These data seem to show that intravenous gammaglobulin replacement does not cause depression of the humoral or cellular immune response or an abnormal sensitization to polyclonal IgG.

In conclusion, our data show that intravenous intact gammaglobulin replacement is both safe and effective in the prophylaxis of infection in high-risk, preterm, very low birth-weight infants with low levels of IgG. Further studies are necessary to determine the minimum effective dose of gammaglobulin and to delineate subpopulations of preterm infants at a very high risk of infection. Follow-up of the treated infants is also necessary to rule out the possibility of long-term sequelae.

References

Amer J, Ott E, Ibbott FA, O'Brien D, Kempe H (1963) The effect of monthly gamma-globulin administration on morbidity and mortality from infection in premature infants during the first year of life. Pediatrics 32:4–9

Boyer KM, Papierniak CK, Gadzala RN, Parvin JD, Gotoff SP (1984) Transplacental passage of IgG antibody to group B streptococcus serotype 1a. J Pediatr 104:618–620

Chirico G, Duse M, Chiara A, Rondini G, Ugazio AG (1983a) Neonatal immunodeficiency and new perspective of prophylaxis of infections with intravenous intact gammaglobulin. XVII International Congress of Pediatrics. Proceedings of symposia, vol 2, p 579

Chirico G, Marconi M, Colombo A, Chiara A, Rondini G, Ugazio AG (1983b) Deficiency of neutrophil phagocytosis in premature infants: effect of vitamin E supplementation. Acta Paediatr Scand 72:521–524

Chirico G, Marconi M, DeAmici M, Gasparoni A, Mingrat G, Chiara A, Rondini G, Ugazio AG (1985) Deficiency of neutrophil bactericidal activity in term and preterm infants. A longitudinal study. Biol Neonate 47:125–129

Christensen KK, Christensen P, Bucher HV, Duc G, Kind CH, Mieth D, Müller B, Seger RA (1984) Intravenous administration of human IgG to newborn infants: changes in serum antibody levels to group B streptococci. Eur J Pediatr 143:123–127

Cooper MD, Gathings WE, Kubagawa H (1982) Immaturity of B lymphocytes in the newborn. Riv Ital Pediatr 8:839–847

Corrigan JJ Jr (1981) Hematologic manifestation of congenital infections. Clin Perinatol 8:499–507

Dubowitz LMS, Dubowitz W, Goldberg C (1970) Clinical assessment of gestational age in the newborn infant. J Pediatr 77:1–10

Dwyer JM (1984) Thirty years of supplying the missing link. History of gamma globulin therapy for immunodeficient states. Am J Med 76:46–52

Eisenfeld L, Ermocilla R, Wirtschafter D, Cassady G (1983) Systemic bacterial infections in neonatal deaths. Am J Dis Child 173:645–649

Fischer GW, Weisman LB, Hemming VG, London W, Hunter KW, Bosworth JM, Sever JL, Wilson SR, Curfman BL (1984) Intravenous immunoglobulin in neonatal group B streptococcal disease. Am J Med 76:117–123

Forman ML, Stiehm R (1969) Impaired opsonic activity but normal phagocytosis in low-birth-weight infants. N Engl J Med 23:926–931

Hayward AR, Lydyard PM (1979) B cell function in the newborn. Pediatrics 64:758–764

Hemming VG, Overall JC, Britt MR (1976) Nosocomial infections in a newborn intensive-care unit. Results of forty-one months of surveillance. N Engl J Med 10:1310–1315

Hill HR, Augustine NH, Shigeoka A (1984) Comparative opsonic activity of intravenous gamma globulin preparations for common bacterial pathogens. Am J Med 76:61–66

Hobbs JR, Yeung CY (1968) Serum G-globulin levels in normal, premature, post-mature, and "small-for-dates" newborn babies. Lancet 1:1167–1170

Marks MI, Welch DF (1981) Diagnosis of bacterial infections of the newborn infant. Clin Perinatol 8:537–558

Morell A, Schurch B, Ryser D, Hofer F, Skvaril F, Barandun S (1980) In vivo behaviour of gamma globulin preparations. Vox Sang 38:272–283

Notarangelo LD, Chirico G, Chiara A, Colombo A, Rondini G, Plebani A, Martini A, Ugazio AG (1984) Activity of classical and alternative pathways of complement in preterm and small for gestational age infants. Pediatr Res 18:281–285

Oxelius VA (1984) Immunoglobulin G (IgG) subclasses and human disease. Am J Med 76:7–18

Pilgrim U, Fontanellaz HP, Hitzig WH, Evers G (1975) Normal values of immunoglobulins in premature and in full-term infants, calculated as percentiles. Helv Paediat Acta 30:121–134

Römer J, Morgenthaler JJ, Scherz R, Skvaril F (1982a) Characterization of various immunoglobulin preparations for intravenous application: I. Protein composition and antibody content. Vox Sang 42:62–73

Römer J, Späth PJ, Skvaril F, Nydegger UE (1982b) Characterization of various immunoglobulin preparations for intravenous application: II. Complement activation and binding to *Staphylococcus* protein A. Vox Sang 42:74–80

Sacchi F, Rondini G, Mingrat G, Stronati M, Gancia GP, Marseglia GL, Siccardi AG (1982) Different maturation of neutrophil chemotaxis in term and preterm newborn infants. J Pediatr 101:273–274
Santulli TV, Schullinger JN, Heird WC, Gongaware RD, Wigger J, Barlow B, Blanc WA, Berdon WE (1975) Acute necrotizing enterocolitis in infancy: A review of 64 cases. Pediatrics 55:376–381
Stèen JA (1960) Gammaglobulin in preventing infections in prematures. Arch Pediatr 77:291–296
Stiehm ER (1979) Standard and special human immune serum globulins as therapeutic agents. Pediatrics 63:301–313
Usher RH (1981) The special problems of the premature infant. In: Avery GB (ed) Neonatology pathophysiology and management of the newborn. Lippincott, Philadelphia, p 255
Van Furth R, Leijh PCJ, Klein F (1984) Correlation between opsonic activity for various microorganisms and composition of gammaglobulin preparations for intravenous use. J Infect Dis 149:511–517
Von Muralt G, Sidiropoulos D (1981) Intravenous Ig substitution therapy in the treatment of septicemia in preterm neonates. In: Nydegger VE (ed) Immunohemotherapy. Academic, London, pp 313–324

Prevention of Atopy: Neonatal Aspects

ROBERT N. HAMBURGER and SETH M. ASSER

Pediatric Department, Immunology and Allergy Division/M009D, University of California, San Diego
La Jolla, CA 92093, USA

1 Introduction

The allergy-prone human infant fortunately differs from other small mammals with regard to early sensitization. It has been reported by Jarrett (1977) that ingestion of low antigen doses predisposes to IgE production in rodents. This observation led to the hypothesis that breast-fed human infants "denied access to large doses of food antigens in infancy" would be more prone to IgE sensitization to those foods in a manner similar to that observed in rodents (Gerrard 1979). That this is not the case is evidenced by double-blind or controlled studies revealing a lowered incidence of IgE-mediated allergic disease in breast-fed human infants compared with infants fed with cow's milk formula (Hamburger and Orgel 1976a, b; Matthew et al. 1977; Ziering et al. 1979; Hide and Buyer 1981; Mellon et al. 1983), confirming the earlier clinical observations of Grulee and Sanford (1936) and Glazer and Johnstone (1953). Furthermore, the longer and more exclusive the breast feeding, the lower the incidence of sensitization and allergic disease (Orgel et al. 1975; Orgel and Hamburger 1976; Saarinen et al. 1979; Hamburger et al. 1983), the lower the total IgE level, and the lower the specific IgE and IgG antibody to cow's milk proteins and to egg proteins observed (Hamburger 1981; Hamburger et al. 1983; Hamburger 1984). The issue, however, is not yet resolved, with other studies (Halpern et al. 1973; Ferguson et al. 1983) revealing no difference in the incidence of allergic diseases in infants fed cow's milk or soybean formulas vs breast feeding, and Björksten and Juto (1983) and Björksten and Saarinen (1978) reporting increased cow's milk sensitization in breast-fed babies.

There is no disagreement that allergy and asthma are by far the most frequent chronic diseases of childhood (NIAID Task Force Report 1979). While the mortality has been declining in recent years, there has been an apparent upsurge in the morbidity. This has been discussed in relation to the regulation of the atopic allergic (IgE) immune system (Hamburger 1973, 1976), the genetics or inheritance of atopic diseases (Hamburger 1980, 1982), and the selection of populations or the ability to mount a higher IgE immune response (Hamburger 1973; Orgel et al. 1974).

Immunology of the Neonate
Burgio, Hanson, Ugazio (Eds.)

2 Development of Immunoglobulin E (IgE)

In studies reported in 1971 by Bazaral, Orgel, and Hamburger, the wide variation in developmental patterns of IgE in the 1st year of life were first noted. In studies performed in early 1972 (Hamburger and Orgel 1982), newborn infants were studied for the presence of skin-fixed IgE before and after exchange transfusions. None of the babies had detectable IgE in the skin prior to the exchange. In the absence of maternal-fetal bleed or contamination of the cord blood with maternal blood (O'Connor et al. 1984), cord serum IgE levels averaged 0.2 IU/ml (Bazaral et al. 1971a; Kjellman et al. 1976; Michel et al. 1980; Croner et al. 1982; Hamburger et al. 1983; Asser et al. 1984; Asser and Hamburger 1985). By conventional methods of serum IgE quantitation, no IgE is detectable in normal newborn cord serum (Bazaral and Hamburger 1972). Special techniques are used to obtain the predictive data relating cord levels above 0.5 IU/ml (Michel et al. 1980) to an increased incidence of early onset allergic disease. Cord serum IgE levels greater than 1.3 IU/ml (the geometric mean plus two standard deviations) were associated with a 70% incidence of atopic allergic disease by age 18 months, and most of the rest of the study patients had signs and symptoms of allergy by age 54 months (Croner et al. 1982). This is strong evidence that the development of serum IgE is under genetic control (Hamburger 1982) and linked to familial tendency to develop allergic disease. In twin studies in children (Bazaral et al. 1974), an analysis of variance gave an estimate of 84% due to genetic control, leaving only 16% of the IgE variation due to the environment; but in adult twins the environmental influence increased to over 30%. It was therefore somewhat surprising to observe the impressive decrease in the incidence and severity of allergy in children who had been in our first "prophylaxis of allergy" study (Ziering et al. 1979; Hamburger 1984).

With the development of an objective method (Mellon et al. 1983) to measure adherence to the prophylactic regimen, in our subsequent collaborative study (Hamburger et al. 1983; Hamburger 1984) it appears that the reduction in atopic allergic signs and symptoms may be greater in those with the best compliance with that somewhat arduous regimen. A more recent estimate (Zeiger et al. 1985) of the effect of dietary compliance on the development of food allergy by the age of 1 year was 10/108 (9%), as against 20/57 (35%) in the noncompliant group ($p<0.001$).

3 Prevention or Delay of Onset of Allergic Disease

The ability to recognize that a prophylactic regimen has reduced the "expected" incidence of atopic allergic disease in based upon the observations that both the family history and the basal serum IgE levels are reliable indicators or predictors (reviewed by Hamburger et al. 1983) of future allergic disease in infants and children. The results of eight separate studies indicate that a bilateral history of major allergy in first-degree relatives results in almost one-half of all children having manifestations of major allergic disease before age 3 years. According to

Kjellman (1976b), 42% of children with allergy in both parents will manifest allergy by 18 months of age which is double the incidence seen in children with a unilateral parental history. Furthermore, if both parents have the same allergic disease, the incidence jumps up to 72% of the children developing the same allergy before the age of 5 years. Thus, the observation of only two babies with allergy among the six with skin test and RAST evidence of sensitization at 4 months of age in the first 123 highly atopic families suggests a markedly reduced incidence of allergic disease in the prophylactic study group (Hamburger 1982).

Comparison of the control group and the prophylaxis group regimes (Hamburger et al. 1983) as reviewed by Asser and Hamburger (1985) is as follows:

Desired Action	*Prophylaxis Group*	*Control Group*
Identify at-risk families	By history, examination , serum IgE level, RAST and skin test in mother and father	Same
Prevent intrauterine sensitization (last trimester)	Hypoallergic diet (individualized). Provide calcium supplementation in place of cow's milk. No whole egg and no peanuts in diet. Avoid food binges. Dietician/Nutritionist available	Routine pregnancy diet as recommended American Academy of Obstetrics & Gynecology
Prevent postnatal sensitization ingestants	Continue hypoallergenic diet during lactation. Withhold all foods other than breast milk for first 6 months Nutramigen (casein hydrolysate formula) if supplement is required. Breast feeding strongly encouraged and supported. Add solids after 6 months; cow's milk after 1 year of age; peanuts and eggs after 2 years of age	Routine American Academy of Pediatrics dietary advice (breast feeding encouraged)
Inhalants	Rigorous dust, mite, and mold controls in home. All animals excluded from home. HEPA filter in infant's bedroom (if available). Nothing stored in bedroom. Environmental home visits (nurse)	Routine pediatric advice given
Contactants	Clothing and bedding of cotton, synthetics (no wool). Hypoallergenic creams, lotions, etc.	Routine pediatric advice given
Nonspecific	Discourage parental smoking. Avoid crowds, especially during community viral outbreaks	Routine pediatric advice given

Monitoring	Allergy nurse practitioners on call for consults and home visits	No home visits
	IgG RAST monitor for compliance	Same
	IgE & IgE RAST for sensitization	Same
	Skin tests for sensitization	Same
	Examination by pediatrician at routine intervals and upon request	Same

4 Collaborative Prophylaxis Study Update

The clinical impressions of the physicians caring for the study patients agree with the spontaneous observations of those parents who have older children not raised on the prophylactic regimen. That is, if the study children have any symptoms of allergy, they are much milder and more easily treated than the older siblings' allergies. While the laboratory data and the clinical data has not as yet been correlated and analyzed (it is still being entered into the computer), it is clear that the study patients and the control patients have measureable differences, such as length and exclusivity of breast feeding, maintenance of environmental controls, and maternal and infant adherence to prescribed diets, the latter confirmed by the measurement of antibody by IgG RAST for cow's milk and egg protein. In a small subgroup of study patients, the IgG anti-cow's milk level averaged 9 OD units at 4 months of age and 21 OD units at 1 year. By contrast, the control infants (no prenatal or 1st year prophylaxis) had 43 and 83 OD units at comparable ages. In fact, even at birth the prophylaxis study infants had levels 31% lower (Hamburger 1984) than those in the matched control group. Another group of study infants who were selected for strict breast feeding without cow's milk supplementation were noted at 1 year of age to have an average of 21 OD units of IgG anti-cow's milk antibody. One month after deliberate introduction of dietary cow's milk, their average IgG antibody level was 60 OD units. This is clear evidence that an increase in IgG anti-cow's milk antibody, as measured by the ELISA method, reflected the inclusion of milk or milk protein in the diet. This then additionally became a measure of adherence to the regimen in our prophylaxis study.

A measure of the effectiveness of the prophylaxis regimen was recently reported by Zeiger et al. (1985) by environmental home visits with regard to removal of pets, dust and mold. Improvement by an average of 40% (33%, 46%, and 42%, respectively) was observed in parental performance (compliance), with 59% unchanged or worse on follow-up visit.

One infant, previously reported (Hamburger et al. 1983), provides evidence for another benefit from an allergen-avoidance regimen in infancy. Because of skin testing and RAST measurements at 4 months of age in both control and study patients, an infant was detected with both tests positive to *Dermatophagoi-*

des farinae (house dust mite). An environmental home visit revealed a heavily infested bassinet. This was removed from the home, the room was repeatedly thoroughly cleaned and a new baby's crib was installed. At 1 year of age, both the skin test and the RAST were negative for *D. farinae* indicating adequate regulation of IgE synthesis when the inciting allergen was removed, even in this allergy-prone infant.

Another interesting observation was that in four of six infants with skin test and RAST evidence of sensitization at 4 months of age, no signs or symptoms of allergic disease were detected by the parents, the pediatrician, or the allergists. As in our earlier studies (Bazaral et al. 1971b), this observation tends to confirm our impression that an increase in both total and specific IgE precedes by weeks to months the appearance of allergic signs and symptoms. It raises the question of whether it might not be valuable, and indeed cost effective, to monitor during the 1st year of life total and specific IgE levels in infants selected for the liklehood of a high incidence (by family history and cord IgE level) of major allergic diseases.

An estimate (Zeiger et al. 1985) of the incidence of development of eczema in the prophylaxis infants revealed 6/197 infants at 4 months, 11/185 at 1 year, 6/170 at 2 years and 0/140 at 3 years for a cumulative percentage of about 12. Similarly for allergic rhinitis 0/197 at the 4 months examination, 1/185 at 1 year, 2/170 at 2 years and 7/140 by 3 years for a cumulative percentage of 7.

The appearance of food-specific IgE during the first 3 years of follow-up in over 185 children on the prophylaxis regimen revealed maximal IgE food sensitization at 1 year with egg, milk, and peanut markedly predominating and soy, wheat, corn, and cod fish (in that order) the next most frequent positives. The soy sensitivity (six cases) was the only sensitization to peak in frequency at 2 years of age.

To date, cumulatively less than 20% of the 185 prophylaxis regimen infants have developed food-specific IgE, and less than 16% have inhalant allergen IgE by 3 years of age. The incidence of concurrent allergic disease appears to be lower than the incidence of sensitization (as is usually the case), and thus the reduction due to the prophylaxis regimen appears to be quite remarkable even though a rigorous comparison with the control group is not yet available.

5 Future Therapy and Prevention

As indicated by Samter (1979), it is impossible to predict the future of the practice of allergy. Some have even expressed the fear that with the medications and treatments already at hand there will soon be no need for allergists. Our view (Hamburger 1983) is that "It is extremely difficult when in the midst of hundreds of exciting new scientific reports to discern which will lead to meaningful therapy." With our current understanding of the biology of atopy we can expect major new therapeutic approaches. One example is regulation via Fc receptors, such as peptides derived from the epsilon chain of IgE. Since the original report in *Science* (Hamburger 1975), despite doubt about the proposed mechanism of action, HEPP, the human IgE pentapeptide, has been shown to be safe and effective in hundreds of individuals in many different centers in the USA (Prenner et al. 1984; Saccor et al. 1985), resulting in the recent submission of a New Drug Application to the Food and Drug Administration. In our own experience, each new approach to therapy demands more sophisticated and specialist judgement for optimal management of the more difficult cases. No miracle new drug or biologic regulator is going to cure all cases of any illness. Until we can replace defective genes in genetic diseases or downregulate hyper-IgE production in atopy, rational management to prevent, postpone, or ameliorate atopic allergic diseases in allergy-prone infants will include breast feeding and a hypoallergenic diet and environment. Even in the absence of a family history of atopy, human breast milk is the best food and affords the best protection for human infants. The ideal pediatric management of childrens' diseases is prevention, and the prevention of atopy continues to be our major goal.

Acknowledgements. We thank Immunetech Pharmaceuticals, Mead Johnson and Co., Pharmacia, Inc. and Boehringer Ingelheim for their support; Dagmara Tubelis for technical assistance; Linda Galbreath and Janice Turner for help with the manuscript.

Susan Heller, M.S., Michael Mellon, M.D., Richard O'Connor, M.D. and Robert Zeiger, M.D. are coinvestigators with the authors in the collaborative University of California, San Diego - Kaiser Permanente Prophylaxis of Allergy in Infants Study.

References

Asser SM, Hamburger RN (1985) Atopic sensitization of the fetus and newborn in relation to the development of the human immune system. In: Milunsky A, Freidman EA, Gluck L (eds) Advances in perinatal medicine. Plenum, New York, p 299

Asser SM, Heredia HM, Hamburger RN (1984) Desarrollo y prevencion de alergia atopica. Alergia 31:81

Bazaral M, Hamburger RN (1972) Standardization and stability of immunoglobulin E (IgE). J Allergy Clin Immunol 49:189

Bazaral M, Orgel HA, Hamburger RN (1971a) Absence of serum immunoglobulin E in newborns: Heterogeneity in time of synthesis in infants. Pediatr Res 5:381

Bazaral M, Orgel HA, Hamburger (1971b) IgE levels in normal infants and mothers and an inheritance hypothesis. J Immunol 107:791–795

Bazaral M, Orgel HA, Hamburger RN (1974) Genetics of IgE and allergy: Serum IgE levels in twins. J Allergy Clin Immunol 54:288

Bjorksten B, Juto P (1983) Allergy and cell mediated immunity in infants. In: Businco L (ed) Advances in pediatric allergy. Excerpta Medica, Amsterdam, p 19

Bjorksten F, Saarigen UM (1978) IgE to cow's milk in infants fed breast-milk and formulas. Lancet 3:624–625

Croner S, Kjellman NIM, Eriksson B, Roth A (1973) Development of childhood allergy in infants fed breast, soy or cow's milk. J Allergy Clin Immunol 51:139

Ferguson DM, Horwood LJ, Shannon F (1983) Asthma and infant diet. Arch Dis Child 58:48–51

Gerrard JW (1979) Allergy in breast fed babies to ingredients in breast milk. Ann Allergy 42:69

Glaser J, Johnston DE (1953) Prophylaxis of allergic disease in the newborn. JAMA 153:620

Grulee CG, Stanford NH (1936) The influence of breast and artificial feeding on infantile eczema. J Pediatr 9:223

Halpern SR, Sellars WA, Johnson RB, Anderson DW, Saperstein S, Reisch JR (1973) Development of childhood allergy in infants fed breast, soy or cow's milk. J Allergy Clin Immunol 51:139
Hamburger RN (1973) Niveles de immunoglobulina e herencia a la alergia e immunidad a los helmintos. Boletin Medico de Hospital Infantil de Mexico 20:309
Hamburger RN (1975) Peptide inhibition of the Prausnitz-Kustner reaction. Science 189:389–390
Hamburger RN (1976) IgE and allergy. In: Barness LA (ed) Advances in pediatrics. Year Book Medical Publishers, Chicago, p 95
Hamburger RN (1979) Prophylaxis of atopic asthma in the first two years of life. Allergol et Immunopathol 72:71
Hamburger RN (1980) The genetic implication of positive RAST in low IgE individuals. Immunol Allergy Pract 2:182
Hamburger RN (1981) Development of atopic allergy in children. In: Johansson SGO (ed) Diagnosis and treatment of IgE-mediated diseases. Excerpta Medica, Amsterdam, p 30
Hamburger RN (1982) Immunogenetics of IgE provides predictive value for the development of allergy. Ann Allergy 49:9–11
Hamburger RN (1983) New experimental approaches to the therapy of atopic diseases. Ann Allergy 50:291
Hamburger RN (1984) Diagnosis of food allergies and intolerance in the study of prophylaxis and control groups in infants. Annales Nestle 42:54
Hamburger RN, Heller S, Mellon MH, O'Connor RD (1983) Current status of the clinical and immunologic consequences of a prototype allergic disease prevention program. Ann Allergy 51:281–290
Hamburger RN, Orgel Ha (1976a) The prophylaxis of allergy in infants debate. Pediatr Res 10:387
Hamburger RN, Orgel HA (1976b) Prophylaxis of allergy. West J Med 125:215
Hamburger RN, Orgel HA (1982) Transfused immunoglobulin E in newborns. J Allergy Clin Immunol 70:143–145
Hide DW, Buyer BM (1981) Clinical manifestations of allergy related to breast and cow's milk feeding. Arch Dis Child 56:172–175
Jarrett EEE (1977) Acitvation of IgE regulator mechanisms by transmucosal absorption of antigen. Lancet 2:223–225
Kjellman NIM (1976a) Immunoglobulin E and atopic allergy in childhood. Thesis, Linkoping University, Sweden
Kjellman NIM (1976b) Predictive value of high IgE levels in children. Acta Paediatr Scand 65:465–471
Kjellman NIM, Johansson SGO, Roth A (1976) Serum IgE levels in healthy children quantified by a sandwich technique (PRIST) Clin Allergy 6:51
Matthew DK, Taylor B, Norman AP, Turner MW, Soothill JR (1977) Prevention of eczema. Lancet 1:321–324
Mellon M, Heller S, O'Connor RD, Hamburger RN, Zeiger RS (1983) No increase in cow's milk (CM) sensitization after delayed CM ingestion in infancy. J Allergy Clin Immunol 71:40–46
Michel FB, Bousquet J, Greillier P, Robinet-Levy M, Coulomb Y (1980) Comparison of cord blood immunoglobulin E concentration and maternal allergy for the prediction of atopic diseases in infancy. J Allergy Clin Immunol 65:422
NIAID Task Force Report (1979) Asthma and other allergic diseases. US Department of HEW, Bethesda (NIH publication no. 79)
O'Connor RD, Zeiger RS, Mellon MH, Heller S, Hamburger RN (1984) The occurrence of elevated cord sera IgE levels due to maternal blood admixture. NER Allergy Proc 5:149
Orgel HA, Hamburger RN, Bazaral M, Gorrin H, Groshong T, Lenior M, Miller JR, Wallace W (1975) Development of IgE and allergy in infancy. J Allergy Clin Immunol 56:296
Orgel HA, Lenior MA, Bazaral M (1974) Serum IgG, IgA, IgM and IgE levels and allergy in Filipino children in the United States. J Allergy Clin Immunol 53:213
Prenner B, Rohr A, Cohen G, Sainz C (1984) Preliminary results of a multicenter study of the antiallergy pentapeptide "HEPP." Ann Allergy 52:240
Saarinen UM, Backman A, Kajosaari M, Siimes M (1979) Prolonged breast-feeding as prophylaxis for atopic disease. Lancet 2:163–166
Saccar CL, McDowell LJ, Mansmann HC, Elvey SM, Robertson L, Bloom NR, Martynec DM (1985) Evaluation of HEPP in allergic rhinitis. J Allergy Clin Immunol 75:185
Samter M (1979) Allergy and clinical immunology: Fifty years from now. J Allergy Clin Immunol 64:321

Zeiger RS, Heller S, Mellon M, O'Connor RD, Hamburger RN (1986) Effectiveness of dietary control for preventing allergic disease during infancy. Ann Allergy (in press)
Ziering RS, O'Connor RD, Mellon M, Cook DA, Tomaszewski M, Street DH, Hamburger RN (1979) University of California San Diego prophylaxis in infancy study: An interim report. J Allergy Clin Immunol 63:199

Treatment of Infants with Severe Combined Immunodeficiency by Bone Marrow or Fetal Liver Transplantation

JEAN-LOUIS TOURAINE

Transplantation and Immunobiology Unit, Pavillon P, Hôpital E. Herriot, 69374 Lyon Cedex 08, France

1 Introduction

Congenital severe combined immunodeficiency diseases (SCID) are relatively heterogeneous conditions with a spontaneously fatal outcome due to infections, usually before 1 year of age. Patients with SCID can be completely cured by bone marrow transplantation (BMT), especially from a genotypically identical donor (Gatti et al. 1968; de Koning et al. 1969; Bortin and Rimm 1977).

When no compatible donor is available, the patients can receive either an haplo-identical BMT or a fetal liver transplant (FLT). In the first alternative, the bone marrow must be treated in vitro to eliminate every T lymphocyte able to induce or enhance a graft-versus-host disease (GvHD). In the second alternative, the fetal liver is best used in combination with the thymus from the same donor, and the optimal age for the fetal donor is 8–12 weeks following fecundation. In both cases, the immunological reconstitution develops more slowly than following an HLA-indentical BMT. Complete isolation in sterile "bubbles" is therefore mandatory.

2 Bone Marrow Transplantation in SCID

In our transplant center, seven patients with SCID have been treated by BMT (Table 1). In each case, the donor was an HLA-identical sibling. There was no ABO incompatibility. The bone-marrow blood was drawn from the iliac bones of the donor by small aspirates of 2 ml, under general anesthesia. BMT consisted of intravenous infusion of $1.5–2.5 \times 10^8$ nucleated bone marrow cells per kg of body weight. Three patients received one transplant and four received two transplants a few months apart. Graft "take" was obtained in all seven patients. Two patients died at 1 and 4 months of previous conditions, and five patients are alive, healthy, and cured of their disease. All five patients have full immunological reconstitution, stable for 1–11 years.

One of the two deaths was due to previous BCG infection and the severity of such generalized BCG infections in patients with SCID has already been reported (Touraine et al. 1981). No improvement of the patient's condition was observed despite the attempt at immunological reconstitution associated with anti-

Immunology of the Neonate
Burgio, Hanson, Ugazio (Eds.)

Table 1. Bone marrow transplantation in seven patients with SCID (Lyons, 1974–1985)

Patient	Treatment	Take	GvHD	Reconstitution	Outcome
T [a] (male)	BMT (sister)	+++	+	Full	Complete recovery
C (female)	BMT (sister)	+++	(+)	Full	Complete recovery
M (male)	BMT (sister)	+++	-	Full	Complete recovery
H (female)	BMT (brother)	++	-	Initial	Death (previous BCG infection)
D [b] (male)	BMT (brother)	+++	-	Initial	Death (malabsorption and hypoglycemia)
S (male)	BMT (brother)	+++	-	Full	Complete recovery
C (male)	BMT	+++	-	Full	Complete recovery

BMT, one or two bone marrow transplants per patient
[a] Patient with ADA deficiency
[b] Patient with bare lymphocyte syndrome

TB antibiotics. The other death was the result of malabsorption and hypoglycemia in a patient who had had many episodes of gastroenteritis before BMT (Touraine, 1981).

The five patients with full immunological reconstitution are four boys and one girl. The previously described incidence of complications in the host using female donors (Bortin et al. 1981) was not encountered in this series. One moderate and one very mild GvHD developed in these infants and they recovered spontaneously. The first infant had a GvHD following a blood transfusion performed before the diagnosis of SCID was suspected. A BMT form an HLA-identical sister was performed soon afterwards, resulting in the disappearance of GvHD and of previous infections. A skin rash and a moderate liver involvement were observed after one week, but they disappeared rapidly and spontaneously. A second BMT from the same donor was not followed by any skin manifestation. In this patient, it is believed that the transplanted cells have permitted elimination of the transfused lymphocytes which were responsible for the GvHD. BMT may therefore contribute to the cure of GvHD.

Ideally, the diagnosis of SCID should be made at birth and BMT performed immediately. It is then a very simple procedure which can almost be conducted in the outpatient clinic. BMT should not, however, be denied to patients with even severe infections, provided that antibiotics and other measures to control infection are at least partially effective before BMT. Except for one patient with a family history of SCID, who was treated soon after birth, all patients subjected to BMT in our institution had previously had severe infections, including pneumocystis and candida infections. These infections were cured by antibiotics and immunological reconstitution, without sequelae.

Following BMT, differentiation markers and functions of T lymphocytes developed progressively over the first 3–6 months. OKT8+ cells increased rapidly in peripheral blood, while OKT4+ cells remained relatively limited in number for the initial 3 months. Immunoglobulin levels rose progressively, especially after the 2nd month. Frequently, immunoglobulins (particularly IgG) of restricted heterogeneity were detected during a transient phase, followed by normal heterogeneity. Four to six months were usually necessary before a completely normal spectrum of immunoglobulin classes and subclasses was obtained, at which time antibody production was also normal.

The origin of various cells has been determined using chromosomal markers, especially in BMT across a sex barrier. In every case, T lymphocytes were derived from donor cells. In SCID patients without B lymphocytes, these cells also developed from donor cells, although a little more slowly than T lymphocytes. In SCID patients with B lymphocytes, most of peripheral blood B lymphocytes remained of host origin. No difficulty in antibody production resulted from the dual origin of these cell populations, donor and recipient being HLA A, B, C and D/DR identical. Insufficient studies have been made to ascertain whether or not most SCID patients develop a large population of monocytes and macrophages of donor origin following BMT; PMN leukocytes and erythrocytes remain of host type.

Some SCIDs are the result of primary enzyme deficiency. The best documented and, for the time being, most frequently detectable enzyme defect leading to SCID is adenosine deaminase (ADA) deficiency. It is known that ADA deficiency hinders development of lymphocytes, especially T lymphocytes (Meuwissen et al. 1975), but whether or not other cells or tissues are altered by ADA deficiency can only be answered by prolonged follow-up of patients with ADA deficiency in whom lymphocytes only are replaced by normal cells. In such a male patient treated 8 years ago by BMT from his HLA-identical and ADA-heterozygous sister, we have observed a specific reconstitution of the T-lymphocyte population by cells of donor origin with XX chromosomes. B lymphocytes, PMN leukocytes, erythrocytes and apparently all other cells remain of host type. Only T lymphocytes had increased ADA activity at a heterozygous level (as all the sister's cells). Besides full reconstitution of cellular and humoral immunities, all physiological functions have been normal. In particular, bone and cartilages, which initially showed minimal lesions (Meuwissen et al. 1975), developed normally. The reconstitution of T-cell compartment with ADA-positive cells was sufficient for a complete cure of the consequences of this enzyme deficiency.

Without doubt, BMT with a genotypically HLA-identical sibling is the best and most satisfactory treatment of SCID. It results in rapid reconstitution and complete recovery, and no additional treatment is necessary. When γ-globulins and antibiotics are given before the transplant and during the initial follow-up, they can be discontinued after a few month. In most SCIDs, no conditioning is necessary prior to transplantation. Only GvHD may be feared and, if it occurs, it is mild or moderate, with a favorable outcome in SCID patients treated by HLA-identical BMT.

Recent improvement in separation of bone marrow cells has permitted significant depletion of T lymphocytes, resulting in decreased risk of GvHD. Techniques combining E-rosetting with soybean - agglutinin fractionation or involving T-cell removal with one or several monoclonal antibodies have been used to treat HLA-haplo-identical bone marrow before transplantation to patients with leukemia or SCID. The initial results have generally been encouraging, although a few problems remain incompletely solved, including lower incidence of "take" and occasional persistence of some donor T cells (Reisner et al. 1981; Levinsky et al. 1982; O'Reilly 1982; Pretice et al. 1982). In the latter cases, the use of cyclosporin therapy in the patient may be helpful. When failure to obtain a "take" occurs, despite satisfactory technical conditions, "resistance" to transplantation, possibly related to natural killer (NK) cell activity, may be postulated, especially in BMT from a non-HLA-identical sibling. Such resistance can be overcome either by increasing the number of transplanted cells or by preconditioning the patient with chemotherapy aimed at decreasing the function of NK - and possibly other - cells in SCID patients with normal lymphocyte numbers and present NK activity.

3 Fetal Liver Transplantation in SCID

When no HLA-identical donor was available for BMT, we always resorted to fetal liver and thymus transplantation (FLTT) for the treatment of SCID patients. We have treated eight patients thus (Table 2). Stem cells from the fetal liver can proliferate and differentiate after transplantation in the host, resulting in lymphocyte repopulation and immunological reconstitution. Because of their immaturity, stem cells confronted by allogeneic antigens acquire immunological tolerance to these antigens. When the stem cell suspension is devoid of lymphocytes already involved in T-cell differentiation, no GvHD can occur initially. Following differentiation of some stem cells into T lymphocytes, no or only minimal GvHD occurs, provided that the allogeneic antigens are present during this development.

It is suggested that the immune reconstitution provided by FLTT is enhanced by simultaneous transplantation of thymus from the same donor (Bortin and Salzstein 1969; Pahwa et al. 1977), which provides a syngeneic environment for T-lymphocyte differentiation of transplanted stem cells. Such an FLTT does

Table 2. Fetal liver and thymus transplantation in eight patients with SCID (Lyon 1976–1985)

Patient	Treatment	Take	GvHD	Reconstitution	Outcome
S (male)	FLTT	+++	-	Full	Complete recovery
B [a] (male	FLTT	?	-	No	Death (previous BCG infection)
C (female)	FLTT	+++	-	Full	Complete recovery
F (female)	FLTT	++	++	Partial	Death (meningitis)
M (male)	FLTT	+	+++	Initial	Death (GvHD and septicemia)
M (male)	FLTT	+++	-	Full	Complete recovery
M (male)	FLTT	+	++	Initial	Death (infection with Salmonella and catheter complications)
T (female)	FLTT	++	-	Initial but still in the early phase	Healthy, in isolation

FLTT, 1 to 5 fetal liver and thymus transplants per patient
[a] Patient with ADA deficiency

not appear to result in an increased risk of GvHD (Bortin and Salzstein 1969), at least when the thymus is at an early stage of fetal development.

For the above-mentioned reasons, we have used FLTT from 8- to 13-week-old (gestational age) fetuses, in all eight patients. Thymuses which contained numerous thymocytes, i.e., 13-week-old thymuses, were irradiated with 4000 rads prior to transplantation. The transplant consisted of the intraperitoneal injection of a suspension of fetal liver and thymus cells. A part of the cell suspension was also injected intramuscularly in the first three patients. Only fresh fetal tissues were used and the viability of all cell suspensions was found in every case to exceed 70%, as evaluated by the trypan blue method. The mean number of cells transplanted from an individual fetus was 8.2×10^8 nucleated cells, i. e., 1.3×10^8 per kg of body weight. No attempt at HLA matching of the donor and the recipient was made.

The patients received 1–5 sequential FLTTs. Our initial approach was to repeat FLTT 5 months after the first FLTT only if results were felt to be insufficient. At present, we carry out a systematic series of three FLTTs a few months apart on each patient, a method which does not seem to have major drawbacks, but which increases the probability of graft "take" from at least one of the transplants.

All patients were isolated in a sterile "bubble" and decontaminated as much as possible prior to FLTT; isolation was prolonged until virtually full reconstitution was obtained, no matter how long this process took. One boy was left for 1.5 years in the isolator and one girl for 3.5 years. Adaptation of these young children to their limited environment has been remarkable, and no significant psychological problem has been encountered either during isolation or after their removal from the "bubble."

Four patients died due to previous BCG infection, to GvHD associated with septicemia, to untractable Salmonella infection, and to neurologic sequelae of meningitis, respectively. Four patients are alive and very healthy. Three of them have full immunological reconstitution. They have left hospital and lead completely normal lives over a follow-up of 3–10 years after the transplant. The fourth patient has been transplanted recently, and it is too early to evaluate immunological reconstitution.

Graft "take" could easily be documented by identification of cells with an HLA phenotype different from that of the host. HLA typing also allowed determination of which of the sequential transplants had taken: the first FLTT in two patients, one of the further FLTTs in the others. Moderate GvHD occurred in three patients, transplanted with fetal tissues less than 13 weeks old. Two GvHDs improved under prednisone treatment, but the previously existent infection lead to death. The other GvHD was virtually cured by prednisone, but neurologic sequelae following meningitis led to death more than 3 years after FLTT (Table 2).

Immunological reconstitution has developed progressively, but was much slower than that after HLA-identical BMT. In the patients attaining full reconstitution, it developed over approximately 2 years (Touraine 1983), stressing the need for decontamination and perfectly sterile isolation. If not isolated, patients with SCID treated by FLTT would not survive long enough for the development of immune defenses against microorganisms. After 2 years, in the above-mentioned patients, numbers of T lymphocytes, T-lymphocyte subsets, proliferative responses to various stimuli, and T-cell cytotoxic responses were virtually normal. Immunoglobulin levels, allohemaglutinins and antibody production following vaccination approached normal after 3 years only. IgG were first of restricted heterogeneity, then normal.

The first child had the following HLA phenotype: HLA A33 B14 B47 DR4 DR5. The fetal donor from whom reconstituting cells derived (second donor) was HLA A1 A2 B18 B27 DR1 DR7. No HLA A, B or DR specificity was therefore shared by donor and recipient. By separation of T- and B-lymphocyte populations, followed by HLA typing, it was demonstrated that all T lymphocytes were of donor origin, while B lymphocytes were of host origin. The monocyte population included two varieties of cells: Some monocytes derived from donor cells, other from host cells. Despite complete HLA mismatch at the A, B, and DR loci, T lymphocytes did cooperate with B lymphocytes and monocytes, resulting in antibody production toward thymus dependent antigens. T lymphocytes of donor origin could, in particular, recognize tetanus toxoid antigen presented by host monocytes and help host B lymphocytes for antitetanus toxoid antibody synthesis in vitro (Roncarolo et al. 1984). They also exerted effector functions on various other host cells. Defense against virus infections in vivo was normal and T-cell cytotoxic functions were present in vitro toward a large variety of target cells.

Comparable findings were obtained in the study of the other patients reconstituted by FLT, demonstrating the efficacy of this form of therapy.

From a review of these cases, as well as from other reports, some factors could be defined as favorable for the outcome of the FLT (Table 3).

Table 3. Favorable factors in FLT

Patient
- Early diagnosis, isolation, and treatment
- Lack of infection
- Prolonged isolation in sterile environment

Transplant
- Fresh fetal tissue
- Fetal donor 8–12 weeks of age
- Repeated transplants (2–6)
- No. of nucleated cells $> 10^8$ kg of body weight of recipient
- IP or IV route
- Male donor?
- Syngeneic thymus associated to liver

Role of compatibility?

4 The "Allo + X" Hypothesis

In contrast to predictions (Zinkernagel 1978), we observed that, following the slow reconstitution induced by FLTT, no important restriction of T-cell function was imposed by either donor of host MHC. T-cell cytotoxic responses toward target cells with a variety of MHC-determinants were present. In addition, T cells did cooperate with HLA-mismatched monocytes and B cells in an antibody production system.

Although we could not determine whether donor stem cells differentiated into host or donor thymus, these results suggested that the restriction of T-cell response by the MHC may not be absolute in human transplantation. Several explanations for this lack of significant restriction in long-term human chimeras can be postulated (Touraine 1980): (a) Partial in vivo restriction when more absolute restriction appears to exist in vitro by lack of a very discriminative assay; (b) absolute restriction in short experiments as in the mouse, but a partial one in long-term T-cell reconstitution; (c) restriction partially imposed by the thymus MHC, partially by the lymphoid- cell MHC; (d) absolute restriction in inbred animals, but a partial one in outbred individuals, in which many cross-reactions between MHC determinants have been demonstrated; (e) possibility to circumvent the restriction, e.g., by progressive development of allo + X recognition ("allo + X" hypothesis).

The apparent discrepancy between experimental data and results in patients following allogeneic FLTT does not appear to result from a species difference or from a major difference between inbred and outbred individuals. Even in classic inbred mice, evidence is now accumulating to suggest that restriction is not a systematic, absolute phenomenon, but rather reflects a relative "preference" of T cells for syngeneic cells in several types of interactions (Zinkernagel 1982; E. Klein, personal communication). When recognition of self + X is not proposed to T cells, alternative mechanisms of recognition (possibily of the "allo + X" type) may develop.

The "allo + X" hypothesis (Touraine and Bétuel 1981; Touraine 1983) has been based on the following facts: (a) in most models where MHC restriction has been unequivocally demonstrated, the control experiments with allogeneic target cells show a low, but present, response; (b) in our chimeric patients, T cells become progressively capable of all types of interaction with allogeneic host cells; (c) in circumstances where MHC restriction is not observed MHC products are still necessary, suggesting that the presence of these molecules at the surface of target cells is important even when they are of a different specificity (Lipinski et al. 1979).

In the thymus there may be develoment of T-cell subsets, each of which has a primary recognition structure for a given type of histocompatibility antigen. In normal circumstances only those T cells with recognition for self-histocompatibility antigens are solicited (because the alloantigens are not encountered), induced to proliferate, and develop a full repertoire of recognition of the large variety of antigens in association with self-recognition. In chimeric patients, a given set of other histocompatibility antigens is also continuously presented to T cells. Those T cells with recognition for the given alloantigens are then solicited, expand by proliferation, and develop the repertoire of antigen recognition in association with alloantigen recognition.

The crucial role of class I and class II antigen presentation to T-lymphocyte precursors in the thymus is suggested by the severe immune incompetence of patients with the bare lymphocyte syndrome (Touraine 1981). The lack of expression of either class I or class II HLA antigens in this condition results in the inability of T lymphocytes to develop the repertoire of antigen recognition.

In conclusion, BMT and FLT represent satisfactory therapeutic solutions to SCID. In the near future, gene therapy will be able to correct ADA deficiency. For most other SCIDs, however, the initial gene defect is not known and the only available treatment will still be, for many years, stem cell transplantation.

References

Bortin MM, Gale RP, Rimm AA (1981) Allogeneic bone marrow transplantation for 144 patients with severe aplastic anemia. JAMA 245:1132–1139

Bortin MM, Rimm AA (1977) Severe combined immunodeficiency disease. Characterization of the disease and results of transplantation. JAMA 238:591–600

Bortin MM, Salzstein EC (1969) Graft-versus-host inhibition: Fetal liver and thymus cells to minimize secondary disease. Science 164:316–319

Gatti RA, Allen HD, Meuwissen HJ, Hong R, Good RA (1968) Immunological reconstitution of sex-linked lymphopenic immunological deficiency. Lancet 2:1366–1369

Koning J de, Van Bekkum DW, Dicke KA, Dooren LJ, Van Rood JJ, Radl J (1969) Transplantation of bone marrow cells and fetal thymus in an infant with lymphopenic immunological deficiency. Lancet 1:1223–1227

Levinsky RJ, Davies EJ, Linch D, Goldstone AH, Beverly PCL (1982) Soy-bean lectin fractionation of bone marrow to prevent graft-versus-host disease (GvH) in mismatched transplants. Exp Hematol 10 [Suppl 10] 96

Lipinski M, Fridman WH, Tursz T, Vincent C, Pions D, Fellous M (1979) Absence of allogeneic restriction in human T-cell mediated cytotoxicity to Epstein-Barr virus infected target cells. Demonstration of an HLA-linked control at the effector level. J Exp Med 150:1310

Meuwissen HJ, Pickering RJ, Pollara B, Porter IH (1975) Combined immunodeficiency disease and adenosine deaminase deficiency. A molecular defect. Academic, New York

O'Reilly RJ (1982) Transplantation in the treatment of SCID; Methods and efficacy. In: Primary Immunodeficiency Diseases, Birth Defects, The National Foundation, March of Dimes

Pahwa R, Pahwa S, Good RA, Incefy GS, O'Reilly RJ (1977) Rationale for combined use of fetal liver and thymus for immunological reconstitution in patients with variants of severe combined immunodeficiency. Proc Natl Acad Sci USA 74:3002–3005

Prentice HG, Blacklock HA, Janossy G, Bradstock KF, Skeggs D, Goldstein G, Hoffbrand AV (1982) Use of anti-T-cell monoclonal antibody OKT 3 to prevent acute graft-versus-host disease in allogeneic bone marrow transplantation for acute leukemia. Lancet 1:700–703

Reisner Y, Kapoor N, Pollack MS, Dupont B, Chagant RSK, Good RA, O'Reilly RJ (1981) Transplantation for acute leukemia with HLA-A and -B non-identical parental bone marrow-cells fractionated with soybean agglutinin and shepp red-blood cells. Lancet 2:327–331

Roncarolo MG, Touraine JL, Banchereau J (1984) In vivo and in vitro anti-tetanus toxoid antibody production in patients with HLA-mismatched T-cells following fetal liver and thymus transplantation. In: Griscelli C, Vossen J (eds) Progress in immunodeficiency research and therapy, vol I. Elsevier, Amsterdam, p 179

Touraine JL (1980) Co-operation between thymus and transplanted precursor cells during reconstitution of immunodeficiencies with bone marrow or fetal liver cells. In: Thierfelder S, Rodt H, Kolb HJ (eds) Immunobiology of bone marrow transplantation. Springer, Berlin Heidelberg, New York, p 141

Touraine JL (1981) The bare lymphocyte syndrome: report on the Registry. Lancet 1:319–320
Touraine JL (1983) Bone marrow and fetal liver transplantation in immunodeficiencies and inborn errors of metabolism: lack of significant restriction of T-cell function in long-term chimeras despite HLA-mismatch. Immunol Rev 71:103–121
Touraine JL, Bétuel H (1981) Immunodeficiency diseases and expression of HLA antigens. Hum Immunol 2:147–153
Touraine JL, Philippe N, Bétuel H, Souillet G, Bétend B, Souteyrand P, Schmitt D (1981) GvHR and infectious complications in SCID patients treated by bone marrow or fetal liver and thymus transplantation. In: Touraine JL, Gluckman E, Griscelli C (eds) Bone marrow transplantation in Europe: II. Excerpta Medica, Amsterdam, p 209
Zinkernagel RM (1978) The thymus: Its influence on recognition of "self-major-histocompatibility antigens" by T-cells and consequences for reconstitution of immunodeficiency. Springer Semin Immunopathol 1:405

Summing Up

Robert A Good
801 6TH Street Souree, PO Box 31020, St. Petersburg, FL 33731-8920, USA

The scientific meeting was opened by the chief of the organizing committee, Professor Burgio, who reviewed embryonic development of the immunological functions. Professor Burgio pointed out that long before birth the human being has developed the essence of full immunological capacity, the capacity to distinguish between self and non-self, ability to produce antibodies, to express cell-mediated immunities, and to make immunoglobulins. But because the fetus represents an allograft and must not damage the mother or the mother's cells, and because the mother's cells must not damage the fetus, the immunological functions of each, at least toward one another, must be tightly controlled and often modulated. There must exist, at very minimum, an immunological compromise between mother and fetus.

With the event of birth the need for this tight control is suddenly removed; and the disadvantage of the compromised immunological functions becomes at least temporarily an issue of immunodeficiency. Outside the uterus, in a hostile world, the fetus and newborn is immunocompromised. Since we do not yet understand the basis for the success of pregnancy and the succession of events that leads to assumption of full immunological responsibility, we find there were pressing reasons for the present conference.

We can also identify a real need to describe, define, and analyze the immunological functions of neonatal life and of prematurity in as much detail as possible. Such analyses must be carried out in a context which will assist us to address this transitional period of immuno-compromise most effectively. Thus, a study of the immunological defects of the neonate and the premature baby in the context of mother–infant interactions and in the context of rapid changes that take place as immunological responsibility is established, seemed to Professor Burgio to be an ample basis for this conference.

In my brief introductory statement, I pointed out that the current view of immunity functions as a self-contained and self-controlled system, in which powerful and essentail influences of cell–cell and molecule–cell interaction between T-cell and B-cell populations and the interactions of each with macrophages and other inflammatory cells and their products, while of extreme importance in these considerations, cannot any longer be considered sufficient. Already it is clear that in addition to these tight internal regulations of immunological elements and functions, there are potentially powerfull external controls as well that can be exerted on the immune systems. These may be influences

Immunology of the Neonate
Burgio, Hanson, Ugazio (Eds.)

exerted by the classic endocrinological systems, or directly or indirectly by neuroendocrine controls. Indeed, one can find evidence even for direct influences of neurohumoral controls on immunological functions. These new issues must be considered as we attempt to understand how the immunological functions make the crucial transition from a quiescent and permissive state of the fetus to the state in the neonate or premature where the immune defenses must struggle to cope with a hostile external environment. Thus, life in the sea of microorganisms, which could compromise the integrity of the individual, necessitates a rapid development to full expression of the adaptive immunological function and nonspecific defenses, and this expression must emerge from a tightly controlled immunosuppressed condition.

The study of the immunology of the neonate, then, must be, in large measure, an analysis of a rapidly changing set of interactions of the lymphoid cells, monocytes, lymphokines, positive and negative feedbacks, external hormones, and neurohumoral influences which have to take place during the early hours, days, weeks, and months of postnatal life. To understand the issues of neonatal immunology and to comprehend the special problems of immunology of premature infants fully, we need to have a thorough understanding of the essential interactions of the three major networks of the body: the neurological, the endocrinological, and the immunological networks.

A Degree of Autonomy for the Peripheral Lymphoid Systems

In his keynote address, Osias Stutman presented evidence which has, heretofore, been somewhat latent in a number of earlier analyses. He ascertained that although the thymus is known to play crucial roles in achieving development of the T-cell system and thus exert a crucial influence in establishing and maintaining the entire immunological apparatus, there exists a rather striking autonomy and independence of immunological components once the peripheral lymphoid systems have been established and fully developed. Immunologically competent cells that exist in spleen and lymph nodes are in reality, surprisingly, autonomous. He emphasized the differences that exist in the development of alloreactivity on the one hand, and self-reactivity and self-recognition on the other, in the different compartments of the lymphoid systems. He pointed out that T helper cells are relatively ineffective in the process of self-recognition. Further, he presented evidence that self-recognition is very much attributable, initially, to a small specialized population of T cells. The latter may, indeed, represent primitive post-thymic precursor cells rather than fully differentiated T cells. Whether these cells are immediate post-thymic elements or immediate pre-thymic elements has probably been somewhat controversial. Stutman looks on them as being immediately post-thymic, while others consider them to be pre-thymic in nature. Nonetheless, the cells involved seem to be distributed very much more in lymph nodes than in the spleen. Thus, they reflect already evidence of specialization of the compartments of cells within the peripheral lymphoid systems. Later, the export of these crucial recognition elements to the lymph nodes and the spleen becomes more balanced, and there seem to be cycles of export of cells from the thymus and import of cells to the thymus that are to be influenced by events in the thymus. Whereas the thymus is a major exporter of T cells at 2 days of age, and still is, to a lesser extent, at 20 days in the mouse, it turns out to be at most a minimal exporter of cells, at 200 days, for example in this species. Rather surprisingly, the presence or absence of the adrenal has little influence on whether thymic involution takes place. But, in spite of thymic involution, the peripheral lymphoid compartment, after 200 days, can carry on very much on its own. It proceeds with its own self-recruiting, self-renewing, and self-educating activities, quite aloof from thymic influence.

Our own experiments and those of Onue et al., in Sapporo, Japan, with fully allogenic chimeras, indicate that there are indeed at least two separate pathways for development of T cells. One proceeds, as all now know, through the thymus, and one seems to be entirely or primarily independent of thymic influence. It thus becomes important to visualize an immunity system which possesses self-maintaining, self-renewing, and self-educating capacities, separate from and independent of those which operate within the thymus. This concept of a T-lymphocyte system with at least two separate lines of cellular development, which impacts on issues of immunorestoration and recognition of what is self, also has crucial implications for attempts at cellular engineering, e.g., with bone marrow transplantation. This also pertains to the specificities involved in the cellular immunities and the crucial restrictions of cell–cell interactions in immunological functions.

Cross Talk Between Chromosomes in Development of Specificities in B and T Cells

Cooper then reviewed details of what is now known of the major processes involved in B-cell development. He stressed the roles of the gene rearrangements for establishing specificity of antibodies. He discussed, especially, the rather astonishing events which must proceed in appropriate sequence as light chain genes become rearranged in the proper order. He pointed out how VDJ rearrangements on chromosome 14 must precede the at first sequential and then parallel selection of constant region genes that will be used to make up the immunoglobulin molecule. In immunoglobulin and antibody production the C-region genes are of greatest importance, because it is the constant region genes selected that determine the distribution of the antibodies and also that determine what effector cells are used and the role antibody plays in certain effector cell functions.

The processes of rearrangement of genes and the other processes of somatic cell mutation, e.g., involving D region genes, all take place completely independently of antigenic influences. Only when immunoglobulins are fully expressed, and only when antibody molecules are displayed on the B cell surface, and only after allelic exclusion has taken place, does antigenic drive become a factor in B cell development. The human neonate is born very late in the context of these B cell developments and by birth – even in the case of a premature – surely has already developed an extensive repertoire of specificities to which its lymphoid cells can react. These events are, of course, consequent on the extraordinary genetic rearrangement events which long precede the moment of birth.Cooper pointed out that a truly amazing communication must exist between events taking place on at least three different chromosomes. Since the reorganization and selection of genes to be used on different chromosomes proceed in a very orderly fashion, events following each other in a precise sequence, such communication seems paramount. For example, kappa gene VJ rearrangements reflect events that take place between genes located on chromosome. 2. But their rearrangement always takes place prior to rearrangements of lambda VJ, and C rearrangements on chromosome 22 occur. Rearrangements of genes that determine the specificity of lambda chains do not take place at all unless the kappa gene rearrangements fail. Further, events on chromosome 14 also appear to be integrated with events on chromosomes 22 and 2, and the joining of kappa or lambda rearranged genes with rearranged genes for heavy chains on chromosome 14 must be coordinated.

It must be realized that the rearrangement of genes that account for T-cell specificities, like many rearrangements of genes to establish immunoglobulin specificities, are early events in the history of the T-cell lineages of cells, although the rearrangement of the genes for T-cell receptors seems to follow quite a different set of rules from those that govern establishment of the B-cell repertoire.

There also appears to be a striking hierarchy dictating which of the genes that permit specific recognition of different antigens, are activated first and which can only be activated later.

The Language of Cell-Cell Interactions

Anderson also considered these issues and pointed out that capacity for recognition of certain antigens takes place in ontogeny long before the capacity to recognize other antigens becomes possible. This subject attracted Silverstein's attention years ago. Thus, constraints on the responses which can be exhibited by the lymphoid cells, and interaction of these cells with one another and with the external environment, are a function of developmental processes, which govern recognition capacity for both T and B cells. These are true for recognition of foreign antigens and also for the T and B cells' recognition of one another. These developmental processes that continue antigen recognition have also been largely completed long before the birth of human neonates and do not represent, for either T cells or B cells, serious limitations on the capacity to respond and function during the neonatal period or even early in the life of most premature infants. T-helper and T-suppressor functions, as well as B-cell and monocyte recognition processes, are still other sets of recognitions that must be considered. However, suppressor and helper functions of T lymphocytes seem to be very different with cells of neonates than are the responses to autologous cells, mitogens, and antigens in older people. The importance of such distinct differences in responsiveness of the B cells, T helper cells, mitogen responses of T cells, and the suppressor populations in the neonatal period, as opposed to those existing in the cells of older children or adults, are poorly understood. There seems no question but that lymphocyte responses of neonates are strikingly different from responses of lymphocytes of older children and adults, and these differences include significant differences in the populations of T cells which are present in blood and tissues. Indeed, differences in proportion of T cells in the blood with different T cell surface markers have been recorded. These findings indicate that much further study of the mechanisms underlying the striking differences between neonatal cells and cells of premature infants and their functions and those of older children and adults is needed before more than the superficial differences between infants and older children can be appreciated. One characteristic of circulating lymphocytes of the neonate is, for example, that B-cell function is predominantly developed and geared to IgM production during the neonatal period. This is true whether the production of immunoglobulins is induced by antigens or mitogens. However, even during the neonatal period, responses to some antigens are quite vigorous while responses to other antigens are nonexistent. Responses to most polysaccharide antigens, for example, are still most defective in neonates. Related to this is the fact that IgG_2 subclass responses are minimal in neonates. The function of T helper cells to assist in immunoglobulin production, by contrast, is quite advanced. Similarly, responses to certain mitogens seem less advanced and those to other mitogens seem more advanced. The functions of IL-1 production by accessory cells appear to be advanced even at birth, as are antigen processing and antigen presentation functions of monocytes and macrophages. Thus, membrane-bound IL-1 and the capacity to produce and secrete IL-1 are well developed in the neonate.

However, it appears that the T cell of neonates are still quite different from those of the adult. There are striking differences in surface markers on T-lymphocyte populations in the blood of the neonate as against those in the blood of older children and adults. These differences in T-cell populations could surely account for a number of apparent functional differences or apparent inadequacies of neonatal T cells. Thus, the apparently deficient functional capacities of lymphocytes of the neonate can be attributed at least in part to develomental differences of both B- and T-cell populations. But differences between the macrophages of neonates and older children are not so striking as are those of lymphocytes.

Breast Milk - Once Again a New Light

Lars Å. Hanson discussed crucial roles played by the mammary gland as a producer of antibodies and immunoglobulins. A new and exciting component of his presentation was the description of an apparent role for a low-molecular-weight inhibitor of bacterial attachment which he has found in milk. The low-molecular-weight inhibitors of bacterial attachment described appear to be the equivalent of free-floating receptors for pili of the bacteria. These molecules are widely distributed in the body and are not just present in milk. However, these "anti-pili" receptors may represent an additional good reason for the breast feeding of infants, especially under conditions where the neonate may be exposed through poor sanitation and poor hygiene to high concentrations of microorganisms that are potentially pathogenic to him or her. The mammary gland functions to produce IgA and/or IgM antibodies, either after local stimulation or after systemic stimulation. Hanson found it possible, with monoclonal antibodies, to protect against colonization by bacteria in a rat model. He used organisms pathogenic to the rat and showed clearly a protective hierarchy of antibodies. He also found evidence that a hierarchy of selection of immunoglobulin by the transport or, as he calls it, the secretory mechanism exists. The "secretory" component has a selective preference for IgA. But if IgA is not present, IgM can and will be transported by the secretory component that is produced by the mammary gland epithelium. Anti-idiotypic antibody also was found to influence production of secretory IgA antibodies, just as it influences antibody production by antibodies, of other Ig classes and subclasses.

In an elegant experimental system with mice he also showed that exogenous monoclonal anti-idiotypic antibodies can be used to control IgA antibody production. Thus, the contributions of antibodies of different Ig classes, and even low-molecular-weight receptors for pili can account for some of the survival advantages of breast feeding. In these ways Hanson clearly showed that understanding of the functional responses of the breasts and the contributions of breast milk is still evolving.

Those Pools of Phagocytes and Their Deformability

Hill considered, in impressive depth, the contributions of phagocytic function to the resistance and deficiencies of resistance to microorganisms by neonatal and premature babies. He was particularly concerned with the susceptibility of neonates to infection with group B streptococci. He discussed the fact that babies, who do not have antibodies to group B streptococci, are those most likely to succumb to this infection. However, his evidence indicates that abnormalities of phagocytic cells may contribute greatly to the vulnerability of newborns to this and other pathogens. He found the chemotactic responses of cells of newborns to be deficient. The calcium flux and increases of adenylcyclase are also defective in neonates. Increases in cyclic AMP, which are anticipated upon phagocytosis, are decreased, as are capacities to phosphorylate compounds. However, one of the most impressive defects he encountered in newborn infants was in the capacity to polymerize contractile proteins. He thus found that the phagocytic cells of neonates do not undergo the same impressive changes in deformability as are seen with phagocytic cells of older children and adults. If one adds pentoxifylline, a phosphodiesterase inhibitor, to cells of older children and adults, one sees a striking increase in the deformability of red blood cells. The cyclic AMP of neonatal cells, however, may already be very high, and the capacity to induce changes in deformability by flux of this second messenger appears to be quite defective in cells of neonates.

Hill found it of real importance to consider phagocytic cells, especially granulocytes, in terms of their availability for circulation in the blood and thus for many defenses. The size of the available storage pool and the size of the proliferating pool of phagocytes and precursors may indeed be quite inadequate in neonates. One of the real deficiencies in the neonatal rat he found to be inadequate reserves of cells in a storage pool and deficiencies of potentially replicating cells in the proliferating pool. The latter, of course, reside especially in the bone marrow. Thus, neutropenia is readily induced in neonates that have been subjected to a number of challenges, and this abnormality seems to be a function of the degree of immaturity of the baby. After exposure in vivo to monoclonal antibodies directed toward neutrophils, the cells are caused to rush to the lung and be sequestered there. C3d was found to play a crucial role in the release of neutrophils from the marrow, and this complement component was found to be only minimally chemotactic. However, when a combination of C3d plus streptococcy was given, an immense chemotactic influence on the neutrophils was seen. Thus, Hill admonished, a relatively superficial consideration of the protective influence of neutrophils in neonates and prematures suggests that this defense is quite good. However, more penetrating analysis reveals and extraordinary vulnerability at this period of life, which involves distinct deficiencies in the capacity to replace cells in this compartment at this age. These findings, it seems to me, contribute important new information to be considered as we attempt to understand the defective defense against bacterial infection we know exists in neonates and premature infants. These descriptions, it seems, contribute significantly to understanding of the extraordinary vulnerability of neonates to infection with streptococci of group B, as well as to infections with certain other bacteria.

Defects of the C Amplification System

Cole considered the relation of the C system to neonatal biology. He reflected especially on regulation of the C system in the bodily defense. He reviewed evidence of the very considerable growth of the C components which occurs during late embryonic life and during the neonatal period. Each of the C components which appear in the embryo and are present in the newborn are now known to be products of the fetus, and very little if any of this important amplification system in the baby is derived from the mother by passive transfer. Indeed, most of the C components are a product of the liver cells, and very few are produced by muscle or skin. One potential source of these components has been thought to be the phagocytic cells, especially macrophages and lymphoid cells. Thorbecke, and others with rather elegant methods of studying biosynthetic processes, have implicated macrophages and lymphoid cells in the production of a number of C components. Factors B and C3 are examples of the components that can be produced by cells other than those of the liver. Under certain stresses, the contributions of the phagocytic and lymphoid cells may be quite significant. LPS stimulation, for example, gives rise to a 15- to 30-fold increase in synthesis and release of factor B or C3 by the reticuloendothelial cells, but such stimulation does not give rise to impressive increases in the production of C2. Whereas adult blood mononuclear cells responded very well to such stimuli, cord blood mononuclear cells did not exhibit such responses. By contrast, the mother's peripheral blood mononuclear cells seemed to work very well in this regard in the immediate postpartum period. However, when gamma interferon plus other factors, such as lipid X and lipid Y followed by LPS, was used, a marked increase in synthesis of C components by the cord blood cells was demonstrated. Thus, it appears that it may be possible to use proper stimulation of component production by peripheral blood mononuclear cells to correct some deficiencies of availability of C components that are present in the neonate or premature infant. Some of our own studies seem to be relevant to these finding. With Geng et al., of Professor Onoe's group at Hokkaido University, in Sapporo, Japan, we investigated the source of C components, using a model of full hematopoietic chimerism produced by bone marrow transplantation across major histocompatibility barriers. The marrow transplants were made after first purging donor bone marrow cells of all unwanted T lymphocytes that might mediate GVHD. We then carried out marrow transplants from C-component-sufficient into C-component-deficient strains of mice, or from C-component-deficient into C-component-sufficient mice; the levels of C4 or C5 components always followed the capabilities of C-component production by the liver rather than the capabilities of donor marrow. In this model it might be of interest to ascertain whether stimulation with LPS will yield increases in C components from bone marrow-derived cells and their derivatives if the marrow comes from the C-sufficient strains and is transplanted to C-deficient strains.

Tempo of Adjustment and the Kinetics of Interaction

Dr. Laurenti, of Rome, recognized that consideration of the adequacy of nonspecific defenses in neonates can be a complex issue. He pointed out that perturbation of the system, in attempting to cope with a process, for example, can result in major changes in the ability of the neonate to adapt. For example, studies of chemotaxis revealed that kinetic issues and rates of change may be of extreme importance, and deficiencies of neonates and/or of prematures to defend themselves against many microorganisms may indeed be functions of sluggishness of responses of the nonspecific cellular defenses. He argued that differences in response to cell aggregation or cell depletion can reflect a slowness to replace cells utilized. Like Hill, he stressed deficiencies in the storage or replicating pools of cells. Issues as mundane as adequacy of blood perfusion can also account for major differences, in the ability of neonates to protect themselves or to respond to the influences of trigger molecule generation, for example in the kallikrein system. Thus, hypoxemia or shock, as examples, may produce profound secondary deficiencies in the ability of the cells of the neonate or premature infant to exercise their functions in bodily defense.

NK Cells

Mantovani then considered the contributions of NK cells or large granular lymphocytes (LGL) in neonates. He commented on the similarity of these cells to macrophages and showed how monoclonal antibodies are helping to define the heterogeneity of the killer cell populations and also of macrophages. He has obtained evidence, for example, which indicates that there are defining surface markers or differentiation surface markers distinguishing mononuclear phagocytes from breast milk, pulmonary alveolus, peritoneal exudate, or blood. Phagocytes from each of these sites may exhibit somewhat different surface phenotypes. Bone marrow macrophages are clearly a different population of cells from pulmonary macrophages, for example. After exposure to interferon, it seemed that some of the macrophages, especially those among the peritoneal exudate cells, could be induced to exercise NK-like activity. Even epithelioid cells and giant cells could be shown to function effectively as killer cells after they had been stimulated by interferons. In the same vein, macrophages from the different pools responded very differently to treatment with thromboxanes or prostaglandins. He found that when the large granular lymphocytes were looked at under the aspect of the most critical criterion for their T-cell lineage, namely, whether or not they show rearrangement of the beta chain genes which contribute to the T-cell receptors, rearrangement could be demonstrated in 10 of 12 such cells studied. Therefore, although according to apparently crucial methodology the LGL are somewhat heterogeneous, most of them clearly mark as cells closely related to other cells of the T-cell lineage. These cells are nonetheless very different from the usual populations of T cells. He considers that they may be derived either from very early T cells or from cells of a lineage closely related to T cells, but having differentiation characteristics that relate them also to monocytes. His findings support much more study of cell lines or clones of cells established from LGL cells.

The IL-2 Receptors - A New Kind of Assay

David Nelson gave a very lucid presentation of an elegant investigation in which he studied IL-2 production and IL-2 receptors of the cord blood T lymphocytes. He pointed out that T lymphocytes can produce IL-2, gamma interferon, chemotactic factors, MIF, lymphotoxins, and LIF. These T lymphocytes, upon activation, also produce IL-2 receptors. Supernatants of activated T cells yield relatively large amounts of IL-2 receptors. He developed a monoclonal antibody against the nonbinding part of the IL-2 receptor. This monoclonal antibody thus could recognize determinants distinct from Tac antigen site on the same molecule, and these antibodies did not interfere with the functions of the molecule. For example, the monoclonal antibody did not interfere with IL-2 binding to the Tac receptor. Thus, he could purify these receptors by affinity separation using Sepharose columns to which IL-2 had been bound, and with his monoclonal antibody to the non-Tac part of the receptor he could quantify the IL-2 receptors.

The monoclonal antibody did not inactivate the T cell through the Tac or so-called IL-2 trigger, and it could be used to quantify the free IL-2 receptor. Increases of IL-2 receptor concentration in body fluids could be used as a most consistent marker of lymphocyte activation. The receptor is released by activated T cells even after the T cells have been irradiated. He then showed, with this new technique, that cord blood lymphocytes are capable of developing IL-2 receptors and releasing these receptors into blood or other body fluids.

After the method for measuring free IL-2 receptors had been perfected he used it to quantify the IL-2 receptors in the blood of patients suffering from a variety of malignant diseases. The patients who showed the highest levels of these T-cell receptors or Tac receptors in the blood were patients with adult T-cell leukemias; the next highest levels were found in patients with Sézary cell (cutaneous T-cell) leukemia or lymphoma. The lowest level of these circulating receptors among the groups of patients studied were encountered in patients with multiple myeloma. Intermediate levels were seen in patients with Hodgkin's disease and certain solid tissue malignancies. Next to those with myeloma, patients with CLL also had rather low levels. Nelson's new technique should permit quantitative indication of T-cell activation, infiltration, and participation in bodily defenses. It also gives an opportunity to compare T-cell activities in neonates, older children, and adults, and in a variety of diseases in quantitative terms, because the circulating blood levels of these T-cell receptors seem to be directly related to the numbers of T cells activated. Dr. Nelson's presentation was most impressive and contained truly new information.

Herpes Virus Infections and Defenses Against Them in Neonates

Anthony Hayward, of Denver, expressed appropriate concern about the severe nature of infections during the neonatal period with several of the members of the herpes family viruses. He indicated that HSVI and HSVII are sometimes acquired during the birth process. These infections can produce devastating destruction in both premature infants and newborns. He indicated that, with the availability of effective therapy for some herpes virus infections with acyclovir the opportunity for long-term study of the neonatal hosts who have such infections and otherwise might die of highly destructive HSV infections is now at hand. Activated T cells can be used to grow HSVI, HSVII, and CMV virus. He also showed that specific antibodies against these viruses can be used to prevent infection or terminate viremia. Such antibodies, he found, can also prevent infection of otherwise suitable target cells. Natural killer cells can be shown to be capable of killing cells of humans infected with one of the herpes simplex viruses, and this killing capacity, which is clearly attributable to NK cells, may be amplified by exposure of the cells to either IL-2 or to gamma interferon. Specific cytotoxic T cells from a patient who has developed T-cell-mediated immunity will release soluble mediators when stimulated by a virus, will induce direct lysis of infected cells, or even can specifically repress, immunity responses to the virus. Neonatal cells or cells from neonates that have been infected are much more highly deficient in their ability to generate specific CMI and to generate cytotoxic T lymphocytes than are lymphocyte populations from healthy adults. Neonatal cells are also markedly inadequate, compared with adult cells, in generation of plaque inhibition or lysis of infected cells. Hayward took T cells from infected neonates and from neonates who had been infected and had recovered and exposed them to gamma interferon plus added IL-2; he showed they responded less vigorously than adult cells exposed to any of the herpes viruses. The circulatory cells of the newborn had, indeed, very few cells responsive to herpes virus antigen. For example, 1 in 600 000 PBL of neonates was responsive, as against 1 in 15 000 PBL of older children or adults. Even the neonate infected with HSV had only 1 in 50 000 cells responsive to HSV, whereas a nonimmune older child might have 1:100000 responsive cells. Infection with the virus generated an immune response, so that 1 in 10 000 cells became responsive in symptomatic patients. IL-2 production after proper stimulation in symptomatic virus-infected neonates, and also after recovery from infection with HSV, was quite impressive compared with that in healthy adults. In neonates there also appeared to be a deficiency of mediator production by monocytes during herpes infection. Further, evidence was presented showing that selective interference with responder cells to the CMV virus or herpes viruses occurred in neonates. By contrast, T-cell helper functions were not depressed by CMV infection or herpes infection in neonates, as they were in older children or adults. Thus, Hayward found a number of deficiencies of T-cell functions and of accessory cell functions directed toward herpes viruses to characterize neonates, who of course are susceptible to HSV infections. Together these abnormalities accounted, in an impressive way, for the greatly increased vulnerability of neonates to infections with each of several herpes viruses.

Neonatal Lymphocytes – Different Populations, Different Markers, Different Responses

Professor Burgio, of Pavia, was concerned in his presentation that neonates are impressively compromised in the immunological perspective. He insisted that the compromise, which can now be quantified to some degree, may explain the increases susceptibility of newborns to many different bacterial, fungal, and viral infections. He mused about the relation these immunodeficiencies might have to potential hazards consequent on immunological reactions that the newborn might have if he or she reacted to maternal antigens. During the period of passive transfer of immunoglobulin, prior to the development of capacity to produce IgG, hazardous immunological reactions to the immunoglobulins transferred from the mother might occur were the infant's immunological reactions not pretty tightly controlled. Thus, he concluded that the immunological deficiencies of the neonate might be a reflection of an essential protective influence of a switch-off process. Reaction could be particularly a problem if T-cell immunity functions that had been shut off were permitted to function without modulation. But the increased susceptibility to viral, fungal, coliform, and group B streptococci which we see may also be the consequence of temporary persistence of this essential down-regulation of immunity attributable to such a shut-off process. He found neonates to have low total T-cell numbers, decreased numbers of OKT3 T cells, and an increased number of OKT8 T cells. In addition, E-rosetting cells were somewhat low in number. Proliferative responses to Con-A and PHA stimulation were, however, not substantially different from those of adults. IL-2 production was also quite normal, and no substantial differences in overall function of T cells could be found in the neonates. T-cell subpopulations, as revealed by surface phenotype analysis, showed OKT4 numbers to be equal to those of adults. An OKT4 plus OKT8 double marker subpopulation of lymphocytes, however, made up 12%–25% of lymphocytes, and this number greatly exceeded that present in adults. The proportion of cells bearing the T10 and T9 markers was also significantly increased in the cord blood. Indeed, as many as 75% of the peripheral blood lymphocytes had these early T-cell markers, which signifies that the peripheral blood T cells in neonates are really quite immature compared with T cells of older children and adults. Cells that have receptors of PNA were strongly represented among peripheral blood cells of neonates, as were cells with TDT markers. The OKT10-marked T cells progressively declined until they disappeared from peripheral blood cells by the age of 6–8 years. Further, the number of OKT4 and OKT8 cells often exceeded the number of OKT3 cells in the blood of neonates. Helper T-cell functions were found to be low, which might be due to the presence of fewer helper T cells or to the influence of active suppression of helper cells. thus, T cells of neonates regularly have surfaces with receptors quite different from those of adults. Many of the Tac+ T cells of neonates were found to have low-affinity Tac receptors for IL-2. Even thought they express Tac antigen, many of them do not seem to be able to respond to IL-2 with the usual vigorous proliferation seen with Tac+ cells in older children and adults. Thus, according to the research of Professor Burgio and his colleagues, the distribution of lymphocyte phenotypes, as defined by surface antigenic expression among lymphocytes from cord blood, is very different from that of older children and adults. Further, the babies appeared to have some lymphocytes suggestive of maternal lymphocytes and also lymphocytes that inhibited functions of maternal lymphocytes that might invade the baby through the placenta. Maternal immunocompetent cells might, thus, attack the baby unless a mechanism were at hand to provide protection against this. The evidence of activation of T cells in cord blood was incredibly frequent, compared with that seen in peripheral lymphocytes of adults. All these characteristics that set the cord blood lymphocytes apart suggest that much traffic does, indeed, occur in the neonate from thymus to peripheral blood and that mothers' cells get into the infant and the infant must be able to shut down cell functions, including its own cell functions, if damage to the developing fetus is to be avoided. The very fact that babies with severe combined immunodeficiency (SCID) often possess impressive numbers of lymphocytes of maternal cell origin is compatible with these findings. Such cells may contribute to the chronic and progressively destructive GVHD in children with SCID. These findings indicate that such controls may be crucial to the developing fetus. It is quite possible that they do indeed depend on functions that continue up to neonatal life and that they themselves involve immunological mechanisms. Such controls then carried over from an in utero existence may contribute to some of the unique immunological characteristics of the neonate. However, in children with SCID we often see a large population of T cells of maternal origin that are not damaging the infant. Why this is so is not well understood. Influence of nutritional factors on these differently deployed populations of lymphoid cells might be crucial to understanding the changes which occur during the transition from in utero to neonatal existence.

NK cells, Leu 7, and HNK-1 cells have been found to be present in lower numbers in neonates than in older children and adults. Why is there an increased number of Tac+ cells in neonates? Perhaps they are attributable to the stimulation afforded by maternal cells in the fetus, which have been shut off by some as yet unknown process in the body.

A Lymphokine Network

Virelizier described a lymphokine network and its importance to the neonatal immunological adjustments. He pointed out that macrophages produce IL-1 or LAF, which can act on T1 cells and also can act to provide helper factors for B-lymphocytes. Activation of T1 cells results in production of IL-2, which in turn acts on T2 cells, inducing proliferation and fostering their differentiation to cytotoxic T cells, helper lymphocytes, and suppressor cells. T2 cells in turn produce interferon, which activates natural killer cells, which in turn can act to modulate the functions of T1 cells. Alternatively, T2 cells influenced by IL-2 can help develop the capacity of other T2 cells to suppress or down-regulate responses of B cell precursors, either by a direct action or indirectly through inhibitory influences on helper T-cell functions.

Virelizier described one patient in whom the capacity to produce gamma interferon and alpha interferon seemed to be dissociated. He also described a patient with SCID, in whom defective secretion of IL-2 appeared to be the definitive lesion. In still another patient a deficient responsiveness to IL-2 was demonstrated, and this deficiency was thought to be the basis for the failure of normal development of a functional T-lymphocyte population. The lymphocyte functions of the patient who failed to produce IL-2 were correctable in vitro by administrations of IL-2 to the patient's cells or in vivo by administration to the patient. Systemic IL-2 also corrected severe immunodeficiencies in this patient. The patient whose cells could not respond normally to IL-2, however, could not be corrected by treatment with IL-2 either in vitro or in vivo. In the patient whose T cells did not produce IL-2, T3 cells were normal in number. Stimulation with PHA, did not, however, yield either IL-2 or gamma interferon production. In the case with IL-2 receptor deficiency numbers of cells possessing T3 were somewhat low. Tac antigen, however, appeared on stimulated cells; IL-1 was produced normally, as were both IL-2 and gamma interferon. However, both the patient's cells and the patient failed to respond with proliferation upon administration of IL-2 to the patient's cells in culture or to the patient. These cases represent dramatic illustrations showing that Virelizier concept of a lymphokine network is clinically useful and that this network is surely of very great importance not only in the neonatal period but also throughout life. The findings are reminiscent of the patients with immunodeficiency described by Geha, in whom IL-1 could be produced but no receptors for IL-1 were present on T lymphocytes or other cells that IL-1 must communicate with. Patients of this latter group, however, have now been treated successfully by bone marrow transplantation after appropriate myeloablation.

Virelizier also described studies of interferon gamma levels in stimulated cord blood cells. He used con-A or anti-T3 antibodies to stimulate the cells to produce gamma interferon. He observed that although the mothers' cells and T cells from normal adults produced large amounts of gamma interferon after such stimuli, the cord blood cells did not. Phorbol myristate acetate, which acted as a potent co-mitogen on normal adult peripheral blood lymphocytes and also increased mitogenic responses to PHA, did not stimulate cord lymphocytes to respond in the same way. Further, the cord cells, unlike other populations of lymphocytes studied, were not induced to produce IFN gamma by treatment with either PHA or phorbol. By contrast, cord blood lymphocytes produced gamma interferon responses comparable to adult responses if the stimulation given was staphylococcal enterotoxin A. Virelizier then performed a crucial experiment. He exposed cord cells to irradiation and later to PHA stimulation. After 100R irradiation to the cells, capacity to respond by IFN gamma production in vitro appeared. After this treatment, gamma interferon could be produced quite normally by the cord blood cells. He further discovered that if he simply incubated the cord blood lymphocytes at 37°C for 24 h the deficiency of IFN gamma responses that characterized the nonincubated cord blood cells was corrected. Thus, the inability of cord blood cells to produce gamma interferon was not attributable to a deficiency of interferon-producing cells, but rather to some inhibitory influence, probably an influence exerted by a putative suppressor T lymphocyte or some other process susceptible to low-dose irradiation. But efforts to find a suppressor cell have not been rewarding. Adherent cells did not account for this suppression. Further, when a cord blood lymphocyte preparation was divided into two, and one-half of the suspension was irradiated prior to PHA stimulation, the newly responsive cord lymphocyte response was not inhibited by a suppressive action of the nonirradiated half of the cord cells. Indomethacin treatment, however, corrected the failure of cord blood cells to produce gamma interferon on stimulation with PHA. By contrast, neither IL-1 nor IL-2 treatment corrected this malfunction. Virelizier's final interpretation was that cord blood cells are inordinately susceptible to down-regulating effects of PGE_2 on gamma interferon production.

Thus, simply a variation in the sensitivity of cord blood lymphocytes to a physiological regulator greatly altered the responsiveness of the lymphocytes and increased the suppressive influence of the mediator. Virelizier then was able to shown that in patients with rheumatoid arthritis there were abnormalities of lymphocyte function attributable to increased sensitivity to the same immunoregulatory influence.

When cells were cultured in vitro over several days they lost their DR positivity. The DR antigens on the surface of lymphoid cells appeared to be actively modulated, and this could be influenced by gamma interferon. Monocytes were lost from the culture in direct relation to the loss of DR expression by the lymphoid cells. Virelizier concluded that gamma interferon, DR expression, and perhaps influences of monocytes and quantitative changes in response to mediators such as PGE_2 might play important roles in regulating the immunological functions of T cells, especially in the neonate.

These studies illustrated, for the first time in the analysis of neonatal lymphocyte physiology, that an extraordinary network of interactions, which involves several cell populations, the production of lymphokines and interactions of cells with the lymphokines, and even sensitivity to certain physiologic mediators may account for apparently defective functions of the cells of the neonate. A corollary, of course, would be that abnormalities of these lymphokine-cell interactions may represent important bases of genetically determined immunodeficiency diseases. The latter may be expressed in the neonatal period, but they then also continue to be expressed throughout life, unless corrected, for example, by bone marrow transplantation.

Clinical Trials of IV IgG for High-Risk Infants

Chirico reported on the intravenous administration of preparations of IgG to high-risk neonates. He reviewed the transient, and often profound, defects of production and levels of IgG and of antibodies in neonates, especially neonates who are also small prematures. He reviewed, extensively, the different chemical and physical modifications that have been used, to date, to prepare gamma globulin for intravenous use. He also reviewed the antibody content of some of the most recent and most promising preparations of immunoglobulins for IV use. He considered neonates with a birth weight lower than 1500 g who required intensive care, especially those requiring assisted ventilation, to be infants with a high risk of infection. He compared babies in this group, given IV IgG 0.5 g/kg/per week with those not given IV IgG. The numbers were quite large – 68 in the treated group and 65 in the control group. He found that the IV IgG preparations significantly reduces the number of infections and the number of deaths in both groups. IV IgG therapy also influenced the number of infections diagnosed. The groups were well matched for Apgar scores, weeks of gestation, and hyaline membrane disease. The occurrence and outcome of infections were clearly beneficially influenced by IV IgG. The IV IgG injections did not cause AIDS or hepatitis B. They produced only mild anaphylactoid reactions, which Chirico attributed to the dimers present, or small amounts of circulating immune complexes in the preparation of gamma globulin used. These results certainly indicate that IV IgG should be cautiously explored as a means of treating or preventing disease in prematures and newborn infants at high risk of infection.

Epilogue

It can be concluded that increasing knowledge of the immunological systems, new understanding of the lymphoid apparatus, appreciation of the network of cellular interactions and definition of the lymphokine - cell receptor and cell modulation networks offer increasing opportunities to analyze the bases of the well-recognized differences in immunological function between prematures and neonates on the one hand, and older children and adults on the other. These accelerating analyses also serve to define new bases for understanding the primary immunodeficiency diseases when the abnormalities persist beyond the neonatal period. Differences in lymphocyte populations, in lymphocyte immunoregulation, in the biologic amplification functions, and in effector functions all can now be shown to distinguish premature infants and neonates.

We have progressed far along the way toward explaining the elevated susceptibility to infection and disease which characterize neonates. However, the feeling remains that although we can already apply some of this new knowledge to lessen the risks for high-risk babies, and can certainly identify some new immunodeficiency diseases in this context, much still remains to be done. Already some of the transient or acute immunodeficiencies can be cured by orthomolecular pharmacology. Some of the more persistent can be permanently cured by marrow transplantation. But this is only the barest beginning. Studies of immunology of neonates and premature infants surely must continue until the minimal essential modulations, possibilities for prevention of diseases, and effective treatments have all been defined. Much important work has been done. Numerous discoveries have been made and many important analyses have been accomplished. Yet, even more remains to be discovered, and many adventures lie ahead as we continue the efforts to understand all the differences between the defenses of premature and neonatal infants and those of older children and adults. It will be through these additional analyses that we will understand better the immunological perturbations that occur with or underlie disease and the immunological peculiarities of prematures and neonates, enabling us to make the adjustments necessary to treat and/or prevent diseases in this crucial phase of life.

Subject Index

MIX
Papier aus verantwortungsvollen Quellen
Paper from responsible sources
FSC® C105338

If you have any concerns about our products,
you can contact us on
ProductSafety@springernature.com

In case Publisher is established outside the EU,
the EU authorized representative is:
Springer Nature Customer Service Center GmbH
Europaplatz 3, 69115 Heidelberg, Germany

Printed by Libri Plureos GmbH
in Hamburg, Germany